# Extreme Weather and Climate Change

**Recent Titles in the**
**CONTEMPORARY WORLD ISSUES**
**Series**

*Campus Free Speech*
Lori Cox Han and Jerry Price

*Global Terrorism*
Steven J. Childs

*Food Insecurity*
William D. Schanbacher and Whitney Fung Uy

*Conspiracy Theories*
Jeffrey B. Webb

*Advertising in America*
Danielle Sarver Coombs

*Cyber Warfare*
Paul J. Springer

*Extremism in the Police*
Carla Lewandowski and Jeff Bumgarner

*The U.S. Criminal Justice System*
Sarah Koon-Magnin and Ryan J. Williams

*Women and Girls in STEM Fields*
Heather Burns Page

*Remote Learning and Distance Education*
William H. Pruden III

*Religious Belief and Science*
Glenn H. Utter

*Child Welfare in America*
Yvonne Vissing

*Water in the West*
Jacqueline Vaughn

*Teen Runaways in America*
Michele Wakin

*Electronic Waste*
Josh Lepawsky

# Extreme Weather and Climate Change

## *A Reference Handbook*

Mariangelica Groves

BLOOMSBURY ACADEMIC
NEW YORK • LONDON • OXFORD • NEW DELHI • SYDNEY

BLOOMSBURY ACADEMIC
Bloomsbury Publishing Inc, 1385 Broadway, New York, NY 10018, USA
Bloomsbury Publishing Plc, 50 Bedford Square, London, WC1B 3DP, UK
Bloomsbury Publishing Ireland, 29 Earlsfort Terrace, Dublin 2, D02 AY28, Ireland

BLOOMSBURY, BLOOMSBURY ACADEMIC and the Diana logo are trademarks of Bloomsbury Publishing Plc

First published in the United States of America 2025



Cover image: Satoshi Kina/Adobe Stock



A catalog record for this book is available from the Library of Congress

ISBN: HB: 978-1-4408-7994-4
ePDF: 978-1-4408-7995-1
eBook: 979-8-7651-1655-5

Series: Contemporary World Issues

Typeset by Integra Software Services Pvt. Ltd.
Printed and bound in the United States of America

For product safety related questions contact productsafety@bloomsbury.com.

# Contents

# Preface

Extreme weather events, such as heatwaves, floods, fires, and hurricanes, have become increasingly common. Scientists now rely on climate records dating back millions of years (found in nature and through instrumental measurements) along with powerful computer models. These tools allow them to assess the likelihood of modern events occurring due to the additional greenhouse gases introduced into the atmosphere by human activity—especially the burning of fossil fuels.

This book serves as an introduction to the science of global warming, which has led to climate change and extreme weather phenomena. It also delves into the challenges associated with preventing further temperature rise. In the following chapters, we explore the current state of the climate, compare it to historical climate changes, and examine the evidence supporting scientific conclusions. Additionally, we discuss attribution science—a statistical approach used to study the influence of anthropogenic global warming on extreme events.

Chapter 1 guides readers through paleoclimatology methods, which decipher past temperature and precipitation extremes influenced by factors like wind patterns, ocean currents, elevation, and Earth's position relative to the sun. We also analyze some of the most extreme events, including heatwaves, droughts, atmospheric rivers, and cyclonic storms during both warm months in the Atlantic and cold winter times in mid-latitudes.

Chapter 2 addresses the problems, controversies, and solutions arising as nations commit to limiting warming to below 1.5 °C (2.7 °F). Meanwhile, they must grapple with the existing increase of 1.2 °C (2 °F) that has already contributed to extreme weather events.

Chapter 3 provides perspectives from various segments of society, shedding light on the challenges we face in mitigating climate change and adapting to its impacts. Readers will get the viewpoint from a top researcher in paleoclimatology and hurricanes, a coastal area manager, a citizen of China researching extreme weather events, a communicator publishing key data for meteorologists and the public on shifts in weather events attributed to global warming, and a member of the Maasai tribe in Tanzania whose pastoral group and work has been affected by climate change.

Chapter 4 is a description of people and organizations related to the topic of extreme weather and climate change. The chapter lists people whose research has provided invaluable information used by modern scientists and policymakers. The chapter also includes advocates and those working to mitigate climate change for vulnerable populations.

Chapter 5 is a data section including figures, tables, and primary documents detailing some of the topics discussed in previous chapters.

Chapter 6 provides an annotated bibliography to guide readers in finding current information on various topics related to extreme weather and climate change. Many of the books suggested will offer an introduction to the concepts discussed in this book. Other books are narratives by top scientists documenting extreme weather events, attribution science developments, and geoengineering efforts to mitigate or adapt to global warming. The chapter also provides websites, documentaries, and podcasts on the subject.

Chapter 7 is a summary in chronological form of the key events that have developed in the understanding of climate change and extreme weather.

# Acknowledgments

I extend my heartfelt gratitude to those who have guided me in acquiring the knowledge I now possess on this subject. During my undergraduate years at Pennsylvania State University, my professors played a crucial role in shaping my understanding. Of utmost significance during that time was Dr. Richard Alley, whose groundbreaking research has left an indelible mark on the field. His commitment to public outreach has been truly profound. I am equally indebted to Dr. Andrea Brunelle, my advisor during my doctoral studies. Her encouragement and meticulous attention to my research made it possible for me to achieve my PhD. Without her support, I would not be a paleoclimatologist. Lastly, I express deep gratitude to my husband, Jeff Groves, whose patience and unwavering love have allowed me to pursue professional projects. My daughters, Eliana and Analina, have also been encouraging and supportive as I continue to build my career.

# 1

# Background and History

We have all seen the images on the television of cities near home and around the world under extreme weather conditions: the people air-lifted off a roof as floodwaters sweep away their house, the line of cars trying to flee an area consumed by fire, the newscasters reporting in front of homes and vehicles buried in several feet of snow or people rummaging through the debris crushed by an intense hurricane or tornado. These images seem increasingly common in our daily news and social media outlets. This chapter explores the science behind the formation of extreme weather events and their connection to anthropogenic (human-caused) global warming (the rise in global average temperature).

It is essential to differentiate weather from climate. Climate refers to the average weather characteristics over thirty years or more. A location's climate primarily results from its relative position under atmospheric wind patterns, the influence of ocean currents, and elevation. Major circulation patterns exist because the Earth is a sphere. Think of the Earth as a beachball with a line drawn horizontally, dividing the ball in half. If we were to expose the beachball (holding it so that the line is horizontal to the ground) to a heat source, the middle of the beachball would get the most heat (since this is the part of the ball closest to that source of heat). The Earth's sphere is geographically divided into two halves. The half that includes the areas from the equator to the North Pole is the Northern Hemisphere. The area of Earth south of the equator is the Southern Hemisphere. The Earth and its continents are not completely vertical

---

Werner Aeschbach (right), of the Hydrospheric Tracers and Proxies group at the Institute of Environmental Physics, University of Heidelberg, Germany, examines an ice core sample from an Austrian glacier before cutting it for argon gas analysis. Variations in the layers of ice can reveal periods of glacier expansion and retreat, giving insight into previous cooling and warming periods. (Sean Gallup/Getty Images)

but are tilted about 23.5 degrees. It is this tilt that exposes various places to the most direct rays of the sun at one part of the year, but not all year round. The Earth revolves around the sun and exposes either the Northern Hemisphere or the Southern Hemisphere more directly to the sun as it changes its position in its revolution. As various latitudes receive direct warming from the sun, these areas heat up more quickly and retain heat.

The atmosphere and the oceans disperse this heat, moving surface ocean currents and atmospheric convection cells from the tropics to the poles. The spin of the Earth and the location of the continents creates various surface wind currents that alternate in direction from the tropics to the poles (Easterlies, Westerlies, Polar Easterlies). The temperature difference from the poles to the tropics creates fast-moving "highways" of air called "jet streams." They form where cold and warm air masses meet and are accelerated by the rotation of the Earth. The more significant the temperature difference, the faster the jet stream.

Elevation also influences a location's climate. Snow-capped mountains give way to lush green valleys. The air at higher elevations is usually colder than the air at lower elevations. The temperature differences between the peaks and valleys can also create vertical wind movement and promote precipitation.

Weather results from pressure and temperature differences on a shorter timescale than climate. Air masses originate over land or oceans and are moved toward various locations by changes in atmospheric pressure that create wind. High-pressure systems stabilize the air and promote clear skies. Low-pressure systems result from the lift of air off the surface of the Earth to higher elevations. The lifted air cools. If there is enough moisture in the air, the moisture can condense into liquid and form clouds and precipitation.

The meteorological information posted on most weather apps includes the hourly temperature, cloud cover, humidity, wind speed, UV index, air quality, and precipitation. Thousands of weather stations collect the data, including those on land, on buoys in the ocean, and on satellites above the planet. They gather real-time information and transmit it to the world's meteorologists (weather scientists) and climatologists (climate scientists).

The Earth's average temperature gives scientists crucial information concerning the amount of energy available in the Earth system for the creation of weather events. An average temperature for the entire planet is a difficult concept to imagine. After all, temperatures on Earth can have a range of nearly 55 °C (99 °F) from a polar location to one in the tropics. Not to mention, the range of temperatures can be tens of degrees on a daily, weekly, or seasonal time frame.

To speak of an "average" temperature may seem like a number with no inherent value. However, a global average temperature is convenient for noting changes to Earth's dynamic systems, including the land, ocean, and atmosphere. One can think of it as an important health measure, like your average body temperature.

How do scientists take all the various temperatures on Earth and come up with a single average temperature for that year? They begin with the temperatures gathered worldwide at multiple stations. The actual readings are then changed into temperature anomalies by calculating the difference between the recorded and long-term average temperatures for each location and date. Government and private organizations that do this have found a common trend: global temperatures are increasing. Many researchers use the GISS Surface Temperature Analysis version 4 (GISTEMP v4) to estimate global surface temperature change. Every month, graphs and tables are updated using the newest data files from meteorological stations and ocean areas, a method developed by James Hansen and others in 2010 and Nathan Lenssen in 2019. Nathan Lenssen is a project Scientist at the National Center for Atmospheric Research (NCAR). His research focuses on understanding climate variability, predictability, and change, while accounting for uncertainties in observational data and biases in climate models. Lenssen has contributed to several significant studies, including improvements to the GISTEMP uncertainty model and research on the predictability of El Niño events.

NASA's Goddard Institute for Space Studies, the National Oceanic and Atmospheric Administration's National Climatic Data Center, and the UK Meteorological Office's Hadley Centre have also kept complete temperature records since 1880.

## What Is Extreme Weather, and How Is It Connected to Climate Change?

Weather changes hourly. Historical records from old ship logs, literature, art, and scientific data collections provide evidence that changes in the weather have occurred many times over. While the weather changes from hour to hour, day to day, and season to season, those who study weather and climate have noticed that some weather events are not within the normal range in their characteristics. These weather events are what are referred to as extreme. The World Meteorological Organization defines extreme weather as follows: "An extreme weather event is rare at a particular place and time of year, with unusual characteristics in magnitude, location, timing, or extent. The characteristics of extreme weather may vary from place to place in an absolute sense" (WMO n.d.).

How do scientists know that a weather event is extreme? How do they know that it is not a repeat of weather phenomena that happens periodically? Extreme weather needs to be put into context. Scientists need to understand how unlikely it is for these events to occur. For a comparison, extreme weather researchers use the knowledge provided by earlier scientific studies that explained the natural changes to long-term weather or climate. The foundation was provided by scientists in the 1800s who used astronomy, math, and geologic evidence to understand natural variability that changed a location's climate. All of these changes happen on very long time scales. For example, a landmass's latitude (degrees north or south) can determine how much solar radiation it receives. Meteorologist Alfred Wegener proposed that continents have slowly (at the rate of the growth of fingernails) moved across the Earth, changing the climate of a location. His 1910 theory is now accepted and referred to as continental drift. The movement of Earth's continents would also change atmospheric and oceanic circulation.

Three major changes in the astronomical position of the Earth in relation to the sun can change the climate of a location for thousands of years and influence the occurrence of ice ages. These are called Milankovitch cycles, after the Serbian scientist Milutin Milankovitch, who published his findings in the 1920s. He coined the first theory—eccentricity. It accounts for a change in the Earth's climate because of the shape of the Earth's path as it orbits around the sun. The gravitational pull from the largest gas planets, Jupiter and Saturn, can cause the shape of the circular path to become elliptical. The variation of the path from circle to ellipse takes nearly 100,000 years. Currently, the Earth's eccentricity is near its most circular. Climate scientists know this is not what is influencing current global warming or the seasonal changes we are experiencing in weather patterns. The rate of change since 1850 is much too fast to match this cycle. The climate has warmed exponentially over the last century.

Obliquity (the Earth's tilt with respect to its orbital plane) is another Milankovitch cycle that can influence the climate on Earth. It is the main driver of Earth's seasons. You may know that the equator is not perpendicular to the plane of the Earth's orbit around the sun. The Earth's tilt is currently at an angle of 23.5 °. Warmer temperatures occur in the Northern Hemisphere when the North Pole is tilted toward the sun as it revolves around it. When the Earth is directly on the opposite side of the sun, the Earth's tilt exposes the Southern Hemisphere more directly to the sun's heat, resulting in the Southern Hemisphere's summer. This is why it is simultaneously summer in the Northern Hemisphere and winter in the Southern Hemisphere.

The angle of the Earth's tilt varies from 22.1 ° to 24.5 ° over 41,000 years, with respect to Earth's orbital plane. The combination of eccentricity and obliquity has caused millennial-scale temperature shifts by as much as 4 °C (7.2 °F) warmer and by as much as 8 °C (14.4 °F) colder. The more extreme the tilt, the more extreme the seasonal temperatures can be. A large tilt can directly expose the poles to the sun and melt ice sheets. Less ice lowers the poles' albedo effect (the reflectivity) as the exposed dark oceans absorb more heat, further warming the planet. When the Earth's tilt is minimal (lessening the exposure of the poles to the sun), ice has more time to form at the poles throughout the Earth's orbit around the sun. More ice increases the albedo effect, decreasing temperatures on Earth.

At 23.5 °, the current tilt of Earth's axis is about halfway between its extremes. Again, the time frame it would take to change our seasons due to obliquity is thousands of years. The amount of change in the tilt of the Earth on an annual or centennial timescale would not have the effects on the ice caps or the atmosphere and oceans that we have seen in the last few decades, especially the extensive warming in the Arctic.

The third Milankovitch cycle that influences very long-term changes in the climate is called axial precession. Not only does the Earth's orbit change from elliptical to circular every 100,000 years, and the tilt of the Earth varies by nearly 2.5 ° every 41,000 years, but the Earth also "wobbles" like a top as it orbits the sun. The wobble moves the axis of the Earth in a circular motion that completes a full circle every 26,000 years. Tidal forces create this wobble. The Earth bulges at the equator as gravitational influences of the sun and moon pull the oceans toward the tropical regions. Axial precession makes seasonal contrasts more extreme as it amplifies the Earth's tilt. Perihelion (when Earth is closest to the sun) occurs during winter in the Northern Hemisphere and summer in the Southern Hemisphere. That is why summer temperatures are more extreme in the Southern Hemisphere and milder in the Northern Hemisphere. However, according to NASA, in about 13,000 years, this will reverse. The wobble of the Earth or the axial precession will expose the Northern Hemisphere to the sun most directly during perihelion, switching temperature extremes to the Northern Hemisphere summer.

All three of the Milankovitch cycles can amplify or dampen the sun's influence at a location on Earth. The most extreme amplification is seen at the poles, where Milankovitch cycles influence the growth or recession of ice sheets. Milankovitch proposed that obliquity (tilt) was most important as it would greatly influence the amount of ice that could form over thousands of years at the poles.

## Paleoclimatology: Looking at the Past to Determine Extremes

Paleoclimatologists are scientists who have studied the changes in climate since the formation of the Earth. Because climate is an average of weather, and weather instruments have not been around for longer than 200 years, these scientists use chemistry and biology to infer what climate was like before the instruments were invented. They compare nature's responses during modern times, from the current instrumental record to the response that nature had hundreds, to hundreds of thousands, to millions of years ago. Fossilized organisms or parts of the natural world that record environmental change are called proxies. Much like a person stands in for another at a meeting when they cannot attend as their "proxy," plants, rocks, fossils, chemical changes, growth rings on trees, and many other records of environmental change act as "proxies" for instrumental data. For example, air trapped as bubbles in ice on frozen areas of the Earth (ice sheets and glaciers) can be extracted to reconstruct atmospheric conditions hundreds of thousands of years ago. Weather patterns can be reconstructed from the chemical and biological records found in sediment deposited for thousands of years at the bottom of lakes and oceans. Other paleoclimatologists use the rings formed within some species of trees to determine weather patterns, including the amount of precipitation that fell in a location during the tree's life span. This practice is called dendrochronology. It is also used to reconstruct water availability driven by weather patterns in an area by analyzing seasonal tree ring growth (more growth = increased moisture). Modern technology helps researchers decipher the changes in the Earth's climate over hundreds of thousands of years using the reconstructions done by paleoclimatologists and computer modeling done by meteorologists or climatologists.

Paleoclimate records support Milankovitch cycles and, along with geologic evidence, pinpoint glacial and interglacial times on Earth. Many people dismiss current climate change with the statement that "the Earth's climate has changed before." This is true, but without the correct time frame, non-climate experts without background knowledge or who have not researched further the natural causes of past changes can promote confusion. As noted earlier, Milankovitch cycles take 10s to 100s of thousands of years to cause climatic changes. The current climate (with anthropogenic emissions of greenhouse gases) is not happening because of natural cycles or natural changes to the Earth's system. They are happening much too quickly and are not aligned with the extreme temperatures amplified by the cycles. In fact, climate models show a cooler average temperature if anthropogenic greenhouse gases are omitted from the inputs. When the models are rerun with the addition of anthropogenic greenhouse gases, they match current observed temperatures.

## The Greenhouse Effect and the Earth's Energy Budget

Climate and atmospheric scientists understood hundreds of years ago that while orbital cycles influenced the timing of the ice ages, there was also a relationship between the physical and chemical properties of the atmosphere that regulated temperatures on Earth. They understood that the atmosphere is composed of gases and that some are very efficient at absorbing outgoing thermal radiation.

The term "greenhouse" gas originated from the idea that the windowpanes in a greenhouse allow the sun's short-wavelength (visible light) energy to enter the greenhouse but trap the outgoing heat (infrared radiation) from escaping back out to the air above the glass. Greenhouse gases absorb thermal energy emitted from Earth and withhold it within the bottom 10 km (6 miles) of our atmosphere. The heat they absorb is reradiated back toward the surface of the Earth. The gases act like the panes in a greenhouse, retaining heat within the atmosphere.

According to NASA, the Earth's energy budget describes the balance between the sun's radiant energy that reaches Earth and the energy that flows from Earth back out to space. About 30 percent of the sun's incoming energy is reflected back to space by objects with high albedo (high reflective properties). These include clouds, atmospheric molecules, tiny, suspended particles in the air called aerosols, and light-colored surfaces like snow, ice sheets, and salt flats.

The Earth's energy budget is directly linked to the Earth's system, which is composed of land surfaces, the ocean, and the atmosphere. The land-ocean-atmosphere system is constantly trying to balance the energy that reaches

Earth from the sun and the energy that flows from Earth back out to space. If changes occur to any part of the Earth's system, Earth's energy budget becomes imbalanced. Earth's temperature increases or decreases to restore an energy balance.

## Global Warming as the Cause of Extreme Weather Events

One of the most significant influences on the Earth's system is a change in the percentage of greenhouse gases in the atmosphere. In the mid-1800s, the Irish scientist John Tyndall identified methane ($CH_4$), water vapor ($H_2O$), and carbon dioxide ($CO_2$) as atmospheric gases that cause the greenhouse effect. In the 1850s, he wrote that any increase in these gases would increase the temperatures in the atmosphere near the surface of the Earth. Eunice Foote, an American scientist who studied heat absorption by $CO_2$, also declared in 1856 that an increase in the gas would result in a warmer planet.

Later, Svante Arrhenius, published an article in 1896 concluding that a reduction in the atmospheric $CO_2$ levels by half would result in a drop in the planet's temperature by 4 °C and 5 °C. Conversely, he added that if those same levels increased by 50 percent, the planet would experience a warming of between 5 °C and 6 °C. Incidentally, paleoclimate proxies (especially the proxies in ice and sediment cores) support the rise and fall of temperatures during increased greenhouse gas from natural factors like volcanic activity. In that same article, Arrhenius identified human industrial activity as the main source of new greenhouse gases in the atmosphere.

## Fossil Fuels: A Nifty Way to Re-use the Sun's Energy

Arrhenius was referring to the early 1700s when human populations became more and more reliant on machines to accomplish everyday simple tasks, heat homes, and power transportation (ships and automobiles). The machines were designed to be powered by the sun's energy but not current solar radiation. The sun's energy that was to be used needed to be extracted from the plant material that absorbed it millions of years ago, the organic material that was transformed into peat, coal, petroleum, and natural gas, also known as fossil fuels.

Fossil fuels are remains of ancient living creatures that absorbed the sun's energy. As ancient plankton (microscopic ocean plants) and terrestrial plants died, the organic material descended to the bottom of the sea and was buried by layers of dirt and rock over millions of years. Heat and pressure from Earth's crust decomposed the organic material into fossil fuels. Oil, natural gas, and coal differ because of the type of fossil, the amount of heat, and the amount of pressure. Plankton decomposes into natural gas and oil, while terrestrial plants become coal.

Fossil fuel energy is produced by combusting or burning ancient plant material that has since turned into coal or organic compounds containing only carbon and hydrogen (hydrocarbons). Petroleum and natural gas are fossil fuels composed of hydrocarbons widely used in industry, heating, and construction. Burning fossil fuels creates heat used to warm homes and power machines. The byproduct of burning fossil fuels and/or mining them produces $CO_2$ and $CH_4$, amplifying the greenhouse effect.

Anthropogenic (human-generated) greenhouse gases add to the amount of infrared radiation (net heat energy) absorbed by naturally produced greenhouse gases already in the atmosphere. The added greenhouse gases join the others in reradiating the extra amount of heat back to the surface, substantially increasing temperatures. The increase in greenhouse gases is compared to having too many glass panes in the greenhouse structure that do not let any heat escape to the outside. Greenhouse gases from burning fossil fuels have continued accumulating in our atmosphere, causing an imbalance in Earth's energy budget. The warmer atmosphere is often called the Human-Enhanced Greenhouse Effect. By the end of 2022, the direct warming influence of human-produced greenhouse gases had risen 49 percent since 1990. The extra warming is referred to as human-induced global warming or anthropogenic global warming. According to the National Oceanic and Atmospheric Administration (NOAA), as of 2023, the Earth's temperature had risen by an average of 0.06 °C (0.11 °F) per decade since 1850. The rate of warming since 1982 was three times as fast at 0.20 °C (0.36 °F) per decade. The warmest year on record since 1850 was 2023, coming in at 1.18 °C (2.12 °F) above the twentieth-century average of 13.9 °C (57.0 °F) and 1.35 °C (2.43 °F) above the preindustrial average (1850–1900). The ten warmest years in the historical record all occurred from 2014 to 2023 (NOAA 2024).

## Greenhouse Gases: Global Warming Potential and Atmospheric Abundance

The Intergovernmental Panel on Climate Change (IPCC) reports that there is scientific consensus that anthropogenic greenhouse gases have caused modern global warming (Eyring et al. 2021). However, not all gases absorb heat with the same efficiency. Their ability to absorb energy (their "radiative efficiency") and how long they stay in the atmosphere (their "lifetime") differs. The Global Warming Potential (GWP) term helps to compare the global warming impacts of different gases. It measures how much energy the emissions of one ton of a gas will absorb over a given period relative to the emissions of one ton of $CO_2$. The larger the GWP, the more that gas warms the Earth compared to $CO_2$ over an average period of 100 years. GWPs provide a standard unit of measure, which helps when total greenhouse gas emissions are added up and policymakers try to find opportunities to reduce emissions across sectors.

Because greenhouse gases are compared to $CO_2$, by definition, $CO_2$ has a GWP of one regardless of the time period used. $CO_2$ emissions cause increases in the atmosphere that last thousands of years. Methane ($CH_4$) has an estimated GWP of approximately twenty-eight times that of $CO_2$ over 100 years and lasts about a decade after it is initially emitted. $CH_4$ absorbs much more energy than $CO_2$. The net effect of the shorter lifetime and higher energy absorption is reflected in the GWP. Nitrous Oxide ($N_2O$) has a GWP 273 times that of $CO_2$ for a 100-year timescale. $N_2O$ emitted today remains in the atmosphere for more than 100 years, on average. The lowest percentages of greenhouse gases belong to the list of chlorofluorocarbons (CFCs), hydrofluorocarbons (HFCs), hydrochlorofluorocarbons (HCFCs), perfluorocarbons (PFCs), or sulfur hexafluoride ($SF_6$). Their GWP is in the thousands or tens of thousands.

The Environmental Protection Agency (EPA) reports the breakdown in the percentage of global anthropogenic greenhouse gas emissions. Emissions of $CO_2$ contribute the most at 78 percent (65 percent from fossil fuels and industrial emissions and 11 percent from forestry and other land uses). Between 1000 CE and the late 1700s, levels of $CO_2$ in the atmosphere averaged between 275 and 290 parts per million by volume (ppmv) of dry air. Since then, fossil fuel use (especially $CO_2$) has increased dramatically as human populations grow and develop infrastructures worldwide that demand high-energy-powered machines. By 1959, $CO_2$ rose to 316 ppmv, and by 2023, $CO_2$ measured 420 ppmv. Modern-day scientists have concluded that rising global temperatures are directly and undoubtedly linked to the addition of greenhouse gases to the atmosphere (most importantly $CO_2$ and $CH_4$) by human activities (U.S. Environmental Protection Agency 2024).

Anthropogenic $CH_4$ contributes to 16 percent of global warming. In the past 250 years, $CH_4$ levels in the atmosphere have more than doubled, rising from

700 parts per billion (ppb) in preindustrial times to almost 1,800 ppb. Isotopic analysis (looking at the different forms of carbon) can distinguish the source. Anthropogenic $CH_4$ is added through agricultural activities (especially rice paddies), leaks from coal mines and natural gas pipelines, fracking (injecting liquid into underground rocks to extract oil or gas), and the clearing of forests by slash (cutting down) and burn methods. Researchers, such as Dr. Nisbet of the Center for Global Change Science, found a rise in $CH_4$ from microbial activity, primarily the microbes in wetlands, rice paddies, and the stomachs of ruminants (mostly cattle). Scientists also hypothesize that the increase in rice-paddy-flooding-events coupled with rising temperatures increases the $CH_4$ output by microbial activity. Some modern satellites are equipped with greenhouse gas sensors. Evidence gathered from satellite scans of the Earth reveals that $CH_4$ production is primarily from the tropics (Nisbet et al. 2016). Atmospheric researchers pinpointed large emissions of $CH_4$ (post-2006) from agricultural fields in China, India, and Southeast Asia, primarily from livestock and flooded rice paddies. Various nitrous oxide gases (like $N_2O$) account for 6 percent of anthropogenic greenhouse gases. Nitrous oxides are added to the atmosphere by agricultural activities, such as fertilizer use and fossil fuel combustion. The remaining percentage of anthropogenic greenhouse gases comes from fluorinated gases (F-gases), which contribute 2 percent to overall emissions. F-gases, including HFCs, PFCs, and $SF_6$, are generated by the industrial processes and refrigeration (IPCC 2014, 465–70).

As mentioned earlier, chemists and physicists like Eunice Foot and Svante Arrhenius warned that adding greenhouse gases would warm Earth's temperatures and create climate shifts. Overall, since 1990, human-produced greenhouse gases have increased the heating power of the atmosphere by 1.11 Watts per square meter, according to the Annual Greenhouse Gas Index (AGGI) monitored by NOAA. Remember that this is an increase in the average watts per square meter of radiative forcing (incoming solar radiation).

The addition of heat to the Earth system has created changes in the characteristics of weather events. Weather events that reach unprecedented size, duration, wind speed, or geographic extent are considered extreme. Extreme weather events are incredibly destructive, costly, and devastating to life on Earth. The IPCC announced that: "Human-induced climate change, including more frequent and intense extreme events, has caused widespread adverse impacts and related losses and damages to nature and people beyond natural climate variability. The rise in weather and climate extremes has led to irreversible impacts as natural and human systems are pushed beyond their ability to adapt" (IPCC 2023).

Meteorologists and climatologists have noticed that extreme weather events continue to set new records. Climatologists who forecast the future use numerical models to simulate the way that climate would respond if physical and/or human activity alters the Earth's atmosphere or oceans. They look at the conditions of the atmosphere (including greenhouse gas content), the oceans (including circulation and temperature of the oceans), and land surface properties (including vegetation cover, land-use change, and ice cover) on Earth.

Rising temperatures increase the amount of water vapor in the atmosphere. When you watch the weather report or if you see a weather map, you will see a large "L" on the map where a low-pressure system has formed. Low-pressure systems are areas with an unstable atmosphere. The air over these locations is lifting off the ground's surface. The lifting air then has a chance to cool (because temperatures decrease with elevation in the first layer of the atmosphere). Cooling air masses can produce condensation, cloud formation, and precipitation. The maps also show areas labeled with a large "H." These areas mark zones where a high-pressure system has formed. High-pressure systems are associated with stability in the atmosphere and clear skies (since air is NOT rising to form clouds). A change in the climate of a region can result from shifts in the location of high-pressure and low-pressure systems.

The movement of a low-pressure system (unstable stormy air) into an above-average warm geographic region can result in record-breaking precipitation totals. This is true due to the simple fact that warm air can hold more moisture. In January 2014, southern England experienced the dual effect described. Increased low-pressure systems formed in the warm Atlantic and moved north. The clouds, filled with a significant amount of water vapor, produced record-breaking rainfall. The warmer atmosphere laden with moisture facilitated heavy downpours, resulting in extreme flooding (Schaller et al. 2016). The floods isolated parts of England, and areas were cut off from the rest of the country when railroads were damaged and large lowland areas were transformed into new lakes.

On the other hand, an altered circulation system may result in fewer low-pressure systems or a shift in location, as wind patterns that would steer these stormy-weather systems are altered. Seasonal rainfall patterns could fail. This explains why some regions may experience drought conditions during a time of year that should have moisture. No rain clouds will form without instability in the atmosphere, especially over dry land. The presence of a high-pressure system can block a low-pressure system. An example of this would be the regions of southwestern Australia, where rainfall has dramatically declined in the last fifty years due to climate change and the shift in global circulation.

Air circulation patterns have shifted because of anthropogenic global warming. The change is referred to as the "dynamic effect." The dynamic effect and the thermodynamic effect (changes in the weather that result from the laws of physics) always occur together. The dynamic effect may vary greatly in strength, location, and severity, making it difficult to simulate in climate models.

It is difficult to understand the changing weather patterns that result from global warming because we often hear about world averages for warming, precipitation, hurricanes, etc. We must look at particular regions and events and understand if they result from a warmer world. This is where attribution studies come in. We will dive into this in the next section.

## Increasing Ocean Heat and Its Role in Extreme Weather Promotion

Ocean temperature anomalies have alerted scientists that there is an overabundance of heat and energy in the Earth's energy budget (the amount of heat and energy entering versus leaving Earth, as discussed earlier). This has to do with the ocean's heat capacity, also known as thermal capacity. It is defined as the amount of heat that needs to be added to increase its temperature. Water has a much higher heat capacity than air (more heat is required in order to raise its temperature than a rock surface, for example). Over a period of the same time, the oceans can absorb the same amount of heat as the land surface without as drastic an increase in temperature. The total amount of heat stored by the oceans is called "ocean heat content." Water temperature measurements reflect the amount of heat in the water at a particular time. Oceans cover over 70 percent of planet Earth. The oceans are very good at absorbing heat from the sun, which keeps our climate somewhat stable as the oceans take many years to change in temperature.

However, 90 percent of global warming is occurring in the ocean, causing the water's internal heat to increase since modern recordkeeping began in 1955. Ocean heat content is measured in joules, a unit of energy, and is usually compared against the 1971 to 2000 average. Scientists at NOAA, Australia's Commonwealth Scientific and Industrial Research Organisation, China's

Institute of Atmospheric Physics, and the Japan Meteorological Agency's Meteorological Research Institute have graphed the increase in heat. The ocean temperatures for the top 700 meters have increased by 360 zettajoules from 1955 to 2021 (the next update will be in 2026). For reference, an increase of one zettajoule ($1 \times 10^{22}$ joules) equals approximately seventeen times the total energy used by all the people on Earth in a year (NASA n.d.).

Heat absorbed by the ocean is moved as part of the hydrologic cycle. The process includes releasing the ocean's heat through the melting of ice shelves, the evaporation of ocean water, and the direct heating of the atmosphere. Because water heats and cools slowly, heat energy in the ocean can warm the planet for decades after it is absorbed. Scientists calculate the heat energy the ocean absorbs and releases to model current and future global climate. The increase in heat energy can increase the amount of water vapor in clouds, manifesting as extreme weather events.

## Attribution Studies: Modeling the Present to Pinpoint the Cause of Extreme Weather Events

Attribution science is a field of research that verifies the likelihood that anthropogenic global warming influences a particular extreme weather event. It focuses on answering a very important question: Is climate change already contributing to its intensity? While early and contemporary climate scientists understood the relationship between the addition of greenhouse gases and an increase in global warming, it wasn't common to quantitatively assess the probability that a particular extreme weather event was made more severe by climate change.

The first attribution study was published in *Nature* in 2004. The paper "Human Contribution to the European Heat Wave of 2003" specifically examined the hottest summer in Europe since around1500 CE and the statistical probability that it was heightened by current greenhouse gas totals (Stott, Stone, and Allen 2004, 610–14). This heatwave will be discussed in more detail later in this chapter.

Although most attribution studies in the early 2000s focused on heat waves, attribution science has expanded and now looks at many types of extreme weather events to determine the contribution of current emissions to their origin and or intensity. In fact, the American Meteorological Society issues a report with attribution science studies for extreme weather events every two years.

The World Weather Attribution (WWA) initiative (founded in 2014) is composed of many climate scientists and meteorologists from around the world, including the College of London in the UK, The Royal Netherlands Meteorological Institute, Le Laboratoire des Sciences du Climat et de l'Environnement in France, Princeton University and the National Center for Atmospheric Research in the United States, Federal Institute of Technology in Switzerland, the Indian Institute of Technology Delhi in India, and climate impact specialists at the international Red Crescent Climate Centre. Scientists at these institutions conduct attribution research to determine the likelihood that a weather event would have had the same characteristics (size, magnitude, intensity, length of time) with and without anthropogenic global warming.

Traditionally, climate studies took months to years to complete and were not of much help or of interest years after an event had passed. Supercomputers and new statistical approaches have shortened the timeline for conclusions. Using extreme event attribution methods, scientists can now quickly (within days to weeks) assess if there is a link between an extreme weather event's characteristics and global warming.

Deciding whether an extreme event is worth an attribution study is often the first consideration. This step is referred to as the analysis trigger. Researchers focus on events that greatly impacted communities or initiated an intense societal discussion. Many times, the events chosen were unprecedented in size, frequency, duration, or magnitude. They resulted in a call for international aid and/or Red Cross relief.

Once an event has been chosen, researchers complete an observational analysis. This step verifies that the data gathered from the extreme event can be reliably compared to the same data from at least the 1950s and preferably back to the 1800s. Local weather agencies verify which data set is most reliable. The data would be compromised if it came from a location where the addition or removal of irrigation, urbanization, or land-use conversion had occurred. The impact on society is not considered as it is impossible to compare from year to year as populations, governments, building materials, and cost of infrastructure change by geography annually. Take the case of flood events from atmospheric rivers in California in 2023 that took lives and cost billions of dollars in damage. The cost and impact are relative to the geography of the event's location. Some places are better prepared or organized for disaster events than others. Therefore, a comparison would not be helpful due to geographical inequalities.

The observational analysis step is particularly important as it can generate two pieces of information: the event's rarity, and changes in global warming trends at the location. Even though the planetary average increase is 1.2 °C, some places have warmed more than others, and some have actually cooled. That is essentially the definition of climate change—a trend away from the average climate, whether warmer or cooler, wetter or drier. The WWA reports that climate change and its effects on weather are so substantial that many extreme events show a clear trend. The trends are statistically calibrated to consider the influence of short-term events such as El Niño Southern Oscillation (a periodic event in the ocean's temperature that can create weather changes).

Climate models add statistical significance to observational data. Using the technology and concepts in weather forecasting, they can show if an event is more or less likely due to global warming. Instead of predicting weather events that typically have a time frame of a few hours to a few weeks, climate models predict how often extreme weather events occur. They must be able to represent the extreme event. Modelers must ask if the statistical distribution of extreme events on the climate model matches the observed data taken over time for a specific location under research.

After the model is run, scientists compute how much more likely or intense the weather event became. The models are run several times using current climate conditions (with anthropogenic greenhouse gas totals). Then, they are rerun in a counter-climate scenario (without current anthropogenic greenhouse gas totals). Scientists count the number of modeled extreme events for each current climate and counter-climate condition. The difference between the two modeled climates tells scientists how much more or less the extremes have become.

WWA researchers and other attribution scientists combine all observational and model estimates. If the observations and models agree, then the confidence that it was caused by climate change due to anthropogenic-induced conditions increases. An example of this step is WWA's assessment of rainfall in the Seine River basin in 2016. The modeled results matched observed rainfall totals with current climate conditions (anthropogenic greenhouse gases). The rainfall totals declined when the models were run in a counter-climate scenario (without anthropogenic greenhouse gases). The matching conditions allowed the WWA scientists to give a reliable attribution statement (Oldenborgh et al. 2016).

Attribution studies are helpful because they consider a population's vulnerability and exposure to a particular event. Consideration is given to the variability of the water cycle and the interaction of our atmosphere and ocean that may cause harm to the public. Researchers calculate the geographical extent to which an event may impact current and future populations. This step is essential as it shows the level at which populations are ready to adapt to climate change-induced extreme weather events. It gives community leaders an idea of what infrastructure needs to be put in place to lessen the impact of those events in the future. For example, the number of casualties from European heatwaves has decreased because communities developed heat plans after the 2003 and 2006 heat waves. This was in response to attribution studies showing the likelihood of the event increasing under current conditions.

Some attribution studies are more accessible than traditional scientific papers and have been produced with easy-to-follow graphics so that the public can better understand the results. Organizations like Climate Central provide graphics for meteorologists, educators, and the public on various topics related to climate change and current weather patterns. These can also be used by government agencies, policymakers, emergency management agencies, and the media.

It is important to note that attribution studies are currently more accurate for larger areas. Because climate models are run on large areas (often 100 miles by 100 miles), the detail needed to pinpoint the cause of an event like a localized rainstorm is limited. Attribution results are more reliable for events covering larger areas, such as heatwaves, large hurricanes or storms, sea level rise, global average temperature, or Arctic Sea ice melt (Cho 2021). For this reason, the extreme weather events covered in this chapter have been statistically modeled to have been more likely due to current conditions (greenhouse gas concentrations at the time).

## Extreme Weather Resulting from Anthropogenic Climate Change

As mentioned, attribution studies work well with weather events that cover large distances, which meteorologists call "synoptic" in scale. Hurricanes, extreme rainfall events, and groups of storms called atmospheric rivers fall under this size prerequisite. Droughts, wildfires, and heatwaves can also occur on a large scale. Even cold spells can result from global warming in some locations, such as those resulting from polar vortexes caused by changes in the Northern Jetstream. This section looks at these phenomena and the destruction they have brought to various regions worldwide.

## Fueled Beasts: Tropical Cyclones

Tropical cyclones, also referred to as hurricanes (in the Atlantic Ocean) and typhoons (in the western North Pacific Ocean), are storms that originate in the warm tropical areas of the world (between the Tropic of Cancer at 23.5 ° north and the Tropic of Capricorn 23.5 ° south). In this area of the world, the sun is always high in the sky, and the energy from the sun delivers a more direct concentration of solar energy per area. This area is also called the Intertropical Convergence Zone (ITCZ). In this region, atmospheric surface winds converge (the easterlies or the Tradewinds). The warm, humid air is pushed to higher elevations in the atmosphere (toward space) as they collide, creating a low-pressure system. The unstable air lifts until it reaches the tropopause (an atmospheric boundary that marks the end of the troposphere and the beginning of the stratosphere). Air at the height of the tropopause (approximately 36,000 feet above the Earth's surface) cools, condenses, and forms clouds. This promotes ideal conditions for the formation of storms and, under the right conditions, tropical cyclones.

Tropical cyclone season begins around May 15 and typically ends around November 30. The peak of cyclonic activity is usually from August 1 to November 10. Only a few storms have formed before May or after December in the past 100 years. The dates make sense, as certain conditions are necessary for forming a tropical storm. Just like a recipe, where the perfect combination of ingredients determines the success of the end product, four critical conditions must exist to create a tropical cyclone. First, environmental conditions must have low vertical wind shear (changes in wind speed and or direction by elevation). The lack of wind shear allows the air to keep lifting and generate tall, large clouds such as those substantial cumulonimbus clouds seen from an airplane window. In other words, the cloud formed from the unstable, humid air is not toppled over or pushed back down by upper atmosphere winds. It is allowed to keep getting taller and taller. Second, a pre-existing disturbance must be present, such as intense convective movement of air forming a group of thunderstorms, a condition referred to as a tropical disturbance. As converging winds gain strength, the resulting clouds can reach altitudes of over 50,000 ft. Third, warm ocean waters with temperatures typically at 26.5 °C (80 °F) to a depth of ~ 250 ft are required. Last, the humidity of the air or the water vapor content must be sufficiently high (Colbert 2022).

If the storm builds and winds reach 23–38 mph, the system is classified as a tropical depression. If winds continue to strengthen and reach between 39 and

73 mph, the storm system is categorized as a tropical storm. After reaching at least 74 mph, the storm is classified as a tropical cyclone. The Saffir-Simpson scale categorizes the velocity of the windspeeds on a scale of 1 to 5 (category one = 74–95 mph to a category five clocking in at >157 mph). Suppose the systems move or originate in higher latitudes and have the characteristics of a midlatitude cyclone. In that case, they are referred to as a subtropical storm. The extension of the northern and southern regions for cyclone activity has also been attributed to the correlating extension of warmer waters into higher (polar) latitudes due to global warming. In 2018, three subtropical storms moved over warmer water, transitioning into hurricanes. The largest number ever recorded at such high latitudes.

Both NOAA and NASA are committed to studying the effects of climate change on tropical cyclones. NOAA's Hurricane Research Division flies directly into hurricanes to sample storm conditions. The data on storms' physical characteristics can be analyzed over time to determine trends. NASA's satellites, equipped with instruments measuring water vapor content, wind speed and direction, and the underlying temperature of the oceans, also provide crucial data to better understand and project hurricane strength.

Earlier in this chapter, we discussed the significant energy increase from 1955 to 2021 in the top 700 m of the ocean (360 zettajoules). Long-term trends in tropical Atlantic warming and the resulting tropical cyclone activity have been attributed to anthropogenic global warming (Mann and Emanuel 2006). Data from ocean buoys and satellite readings have confirmed that sea surface temperatures in the Caribbean have risen at a faster rate from 1980 to 2015 (Taylor et al. 2018, 6785-92) and are projected to keep increasing (Jones 2023).

According to NASA's Jet Propulsion Laboratory research, sophisticated global climate models predict less frequent tropical cyclones. However, the rainfall delivered by tropical cyclones is projected to intensify, and coastal surges may be higher due to rising sea levels. Scientists at NOAA also use computer models to determine changes in hurricane strength, longevity (how long they last), predicted paths, and rainfall production. The increased ocean temperatures, additional humidity in the air, and rising sea levels resulting from human-caused global warming are all variables added to the models. NOAA's climate models agree with those at NASA. Both attribute and project an increase in rainfall from tropical cyclones.

Scientists remind the public that even if the frequency of tropical cyclones stays the same or declines, as the air continues to warm due to global warming, tropical cyclones will hold more water vapor, and rainfall amounts will increase (Colbert 2022). An increase in wind intensity is also projected as a result of increased humidity (from higher evaporation rates) and warming ocean temperatures. Although equal or fewer storms may form, the majority of those storms are predicted to reach a Category 4 or 5 level (Colbert 2022). Another characteristic of concern is the rapid intensification (increase in size and intensity much faster than the historical record) of modern hurricanes.

### *Case Study: Typhoon Hagibis, Japan*

On October 12, 2019, Typhoon Hagibis hit Japan, dumping more than 240 mm of rain on the Tokyo region. No higher rainfall amounts had been recorded since records began in 1976. As of 2022, it was the second-costliest Western Pacific typhoon, leading to nearly 100 deaths and major property loss. Researchers from the Imperial College of London and Oxford University used the exact attribution science developed by the WWA Center to determine whether Typhoon Hagibis had been augmented by anthropogenic global warming. The scientists concluded that Typhoon Hagibis was 67 percent more likely to occur because of human-caused climate change, the warming of the oceans, and the increase in humidity. One of the most interesting parts of this study was that the scientists also figured out how much human-caused climate change had added to the cost of the typhoon. They concluded that $4 billion of the $10 billion in damage caused by rainfall destruction could be attributed to climate change (Li and Otto 2022).

### *Case Study: Hurricane Harvey, Texas, United States*

Hurricane Harvey originated as a low-pressure system on August 13, 2017, off the West Coast of Africa. Near the Cabo Verde Islands, the tropical wave merged with a broad area of low pressure and moved toward the Eastern Caribbean Sea. On Thursday, August 17, Harvey reached tropical storm status. Warnings were issued that afternoon for the countries of Martinique, St. Lucia, Barbados, and St. Vincent and the Grenadines (all southern Caribbean islands just north of South America). On Tuesday morning, August 22, Tropical Storm Harvey moved to the Yucatan Peninsula of Mexico. As the tropical storm moved northwest over the Caribbean, it encountered warmer water and less wind shear, two conditions that strengthened the storms. By Thursday, August 24, Hurricane

Harvey made landfall in south Texas. After landfall, there was a period of rapid intensification. By early Friday evening, Hurricane Harvey was a Category 4 hurricane with sustained winds of 130 mph. The Center for Severe Weather Research clocked wind speeds at Aransas Airport of 145 mph. No one knows the highest wind speeds since most wind stations were destroyed before readings could be gathered. Tens of thousands of south Texas residents lost power for days. Buildings and other property were destroyed. Storm surges of up to 12 ft. were recorded at the Aransas Wildlife Refuge.

Harvey's damage also came from rainfall. The hurricane stalled over southeast Texas for days, and some areas received over 40 inches of rain in just under 48 hours. The North American record was broken at Cedar Bayou in Houston at 51.88 inches. Rivers like the Guadalupe River, the Garcitas, and Coleto Creek broke their banks, flooding southeast Texas. According to Texas State Climatologist John Nielsen-Gammon, Hurricane Harvey produced the most considerable rainfall of any US hurricane on record. Harvey did not stop its course in Texas; other Gulf of Mexico states, like Louisiana, were also under Hurricane Harvey's destructive path as it continued across the coast. It tied Hurricane Katrina as the costliest tropical cyclone, with damages totaling over $125 billion (Halverson 2018).

The rapid intensification and the unprecedented amount of rain from Hurricane Harvey triggered an analysis by scientists curious to know the influence of anthropogenic global warming on the formation, wind speed, and rainfall amounts recorded from the storm. Researchers made several important conclusions. First, they determined that the increase in the surface-ocean heat content led to the extreme rainfall amounts. Hurricane Harvey took considerable amounts of heat out of the overheated ocean, transferred it into the climate system, and converted it to heavy rainfall (Trenberth et al. 2018). Second, human-induced climate change likely increased Hurricane Harvey's total rainfall by at least 19 percent and likely increased the chances of heavy rainfall by a factor of at least 3.5 (Risser and Wehner 2017, 12,463).

Kerry Emanuel, a well-known meteorologist whose expertise includes hurricanes, also conducted an attribution study on Hurricane Harvey. His goal was to create a quantitative assessment of the risk of a hurricane like Harvey. He and his colleagues developed a physics-based risk assessment tool to model the probability of extreme hurricane-delivered precipitation. The results could then give government agencies and disaster management agencies a way to assess hurricane-induced flood risk even if they did not have prior historical records for an area. They estimated the future likelihood of a "Hurricane-Harvey-type" rainfall magnitude using six climate models under current climatic conditions and future climate projections at various greenhouse gas concentrations. They concluded that the probability of Hurricane Harvey's rainfall amounts in Texas was around 1 percent from 1981 to 2000. Using a worst-case high emissions scenario for the amount of greenhouse gases (IPCC AR5 representative concentration pathway 8.5), the probability could increase to 18 percent over the time frame of 2081–2100 (Emanuel 2017).

### *Case Study: Hurricane Maria, Puerto Rico*

Shortly after Hurricane Harvey hit the US Gulf Coast, a mighty hurricane developed in the Caribbean. On September 16, 2017, Hurricane Maria developed as a tropical storm in the Lesser Antilles islands. It, too, intensified quickly and rose to a Category 5 hurricane as it made landfall on the island of Dominica on September 18. As it whipped over the island, maximum sustained winds were recorded at 175 mph (20 km/h). The damage to Dominica was unprecedented. Hurricane Maria had stripped the ground of vegetation, destroyed 25 percent of homes, and left many Dominicans without shelter, food, or water for days. The lasting impacts were especially felt in the agriculture sector, a primary source of income for the country. Saint Croix and neighboring islands suffered extensive flooding, structural damage, and uprooted trees.

On September 20, Hurricane Maria made landfall in Puerto Rico. The people of Puerto Rico suffered from the 155 mph winds, which blasted away their homes, businesses, recreational centers, hospitals, telephone and electrical towers, and tourist-attracting resorts. Hurricane Maria had the greatest 24-hour rain intensities of all storms ever recorded in Puerto Rico (Ramos-Scharrón and Arima 2019). Rainfall totals at the island's center registered at 965 mm (38 inches). The rainfall was 30 percent higher than the previous rainfall record set by a tropical storm in 1985. The quickly saturated ground caused massive landslides. Floods brought sand and saltwater into buildings, adding to property loss. The loss of resources was compounded by a slow relief process and the prior losses from Hurricane Irma, which had hit the island on September 7 (a couple of weeks prior). The island also had the worst electrical blackout in US history, lasting months. For up to ten months, 88 percent of Puerto Ricans (3 million people) remained without power. Estimates suggest the storm killed about 3,000 Puerto Ricans.

Hurricane Maria is the worst natural disaster on record for the Caribbean islands of Dominica, Saint Croix, and Puerto Rico. It also has the distinction of the Atlantic's tenth most intense storm on record. Monetary damages were calculated at $91.61 billion (the third costliest tropical cyclone on record and most costly in Puerto Rican history). With a loss of life of over 4,000 people in the region, Hurricane Maria is the deadliest storm to hit the group of islands since the early 1800s.

As modern hurricanes develop in the Atlantic, they gain strength from rising sea surface temperatures—a major proponent in the intensity of storms of recent years (Emanuel 2017). Attribution studies were conducted to see if global warming amplified Hurricane Maria. When researchers at the Computational Research Division in the Lawrence Berkeley National Laboratory used models that allowed for the use of convection (the process of fluids including air and water to decrease in density and rise when heated, then cool and become denser) in their models, they found that: "Relative to preindustrial conditions, climate change has enhanced the average and extreme rainfall of hurricanes Katrina, Irma, and Maria but did not change tropical cyclone windspeed intensity. In addition, future anthropogenic warming would robustly increase the wind speed and rainfall of eleven of thirteen intense tropical cyclones (of fifteen events sampled globally)" (Patricola and Wehner 2018, 339–46). Researchers from NOAA and the NWS published an attribution study in 2021. The scientists agreed with earlier studies that the monthly average Sea Surface Temperature of 30 °C or 86 °F (0.5 °C or 0.9 °F warmer than average), the high air temperatures (the hottest year on record at the time), and the low vertical wind shear of 4–8 m/s fueled Hurricane Maria (Pasch, Penny, and Berg 2018).

### *Future Hurricane Risk*

After the 2017 season that brought Hurricanes Irma, Harvey, and Maria, attribution studies quantified the relationship between global warming and other tropical storms. They concluded that human-induced warming has increased rainfall associated with large storms in the recent decade, including Hurricanes Irma, Harvey, and Maria in 2017, Florence in 2018, Dorian in 2019, tropical storm Imelda in 2019, Typhoon Hagibis, the extra-active 2020 hurricane season, and Hurricane Ian in 2020, to name a few.

Sea level rise also adds to the destructive power of modern tropical cyclones. Global sea levels are now 203 mm (8 inches) higher than in 1880. They could rise from 1.5 feet to 3 feet or more (depending on the continued input of greenhouse gases into the atmosphere) by 2100. Tropical cyclones raise water levels above the average tide level as they pass. This is referred to as a storm surge. Because cyclones are a low-pressure system over water, the water is lifted much like a vacuum would as it passes over. The counterclockwise spin of cyclones in the Northern Hemisphere (cyclones spin clockwise in the Southern Hemisphere) sweeps water over land. It keeps pushing it forward until the storm passes or loses strength. Storm surges can reach high levels and cover cars and buildings, flooding everything in their midst. The stronger the cyclone, the higher the storm surge. The lower the elevation of the cyclone's contact area, the more possible damage to property and life. The damage from coastal storms has tripled since 1980 along the East and Gulf coasts of the United States (Park 2021).

## Extreme Rainfall

In order to gain an understanding of how current conditions have equated to extreme rainfall totals, it is important to take a closer look at the hydrologic cycle. Rain requires the presence of clouds, which are formed when water or ice evaporates from the Earth's surface or when plants release oxygen and water as a byproduct of photosynthesis (evapotranspiration). The evaporated water rises from the surface of the Earth into the atmosphere in the form of water vapor. As the water vapor cools and condenses, clouds begin to form. The water vapor must have a solid particle, a condensation nucleus, on which to condense. These particles or aerosols can include dust, pollen, water droplets, or ice crystals. The droplets combine and merge inside a cloud through "collision coalescence." As the droplets increase in size, they eventually become too heavy to remain in the cloud and fall to the ground as precipitation, such as rain. This process typically occurs when moist air rises and cools to its dew point, forming water droplets or ice crystals. The dew point is the atmospheric temperature below which water droplets begin to condense and dew can form. The dew point can vary depending on atmospheric pressure and humidity.

The simple process of cloud formation and the amount of rain produced within the cloud is influenced by global warming. Warmer surface temperatures, and the cloud and precipitation formation processes, are fundamental to studying weather patterns and climate dynamics. NASA has used algorithms within forecast models to project the effects of global warming on the amount of precipitation at specific locations. Current climate models indicate rising temperatures will intensify the Earth's water cycle. Warmer surface temperatures will increase evaporation. Increased evaporation will result in more frequent and intense storms in some parts of the world. Areas on the path of increasing storm formation will likely experience higher precipitation and flooding, much like the studies on tropical storms and cyclones discussed earlier. Areas located at a greater distance from storm tracks are likely to experience less precipitation and an increased risk of drought.

### *Case Study: Extreme Rainfall Totals in West Africa*

Even though droughts are increasing in the southern regions of Africa, WWA studies have linked climate change to heightened heavy rainfall, leading to massive flooding events in vulnerable communities in West Africa. This looked to be the case when, in 2022, large regions of West Africa experienced extensive flood events caused by above-average seasonal rainfall and water management issues. The anomalous rainfall began early that year, striking a large area encompassing Mali, Cameroon, Nigeria, and Niger. Those countries experienced the first downpours and consequent flooding in early June. Later that summer, Chad, Cameroon, and Benin were doused by the rains. Nigeria and Chad suffered nearly 800 fatalities combined. Nigeria reported over 3.2 million people affected, with 1.5 million people losing their homes and nearly 3,000 injured. Flooded farmland also threatened food availability.

Scientists from Nigeria, Cameroon, India, the Netherlands, France, Denmark, South Africa, Sweden, the UK, and the United States published peer-reviewed event attribution studies on those extreme rainfall events. The study concluded that for a 7-day maximum rainfall over the lower Niger Basin, models indicated that climate change made the event about twice as likely and approximately 5 percent more intense. For the Lake Chad region, the models concluded that climate change made the event nearly 80 times more likely and 20 percent more intense (Hounkpè et al. 2016). The observations were higher than what the model outputs generated.

Events like these have become more common as the climate warms. The average time between 7-day seasonal maximum events like those recorded in 2022 in the Lake Chad Basin is one in ten years and one in five years for the lower Niger Basin. Climate models predict increased rainfall intensity when the Northern Hemisphere is more directly in the path of the sun's heat during the Northern Hemisphere summer. This creates a lifting of air around the Tropic of

Cancer, leaving a gap for southern air to fill. The southern air is moist as it forms over the oceans of the Pacific, the Indian Ocean, and the South Atlantic. The moist air moves northward from May through September, bringing rain (the wet monsoon). This is what, regionally, people call the North American Monsoon, the Asian Monsoon, or the African Monsoon. The winds change direction as the Earth revolves around the sun and expose the Southern Hemisphere to direct radiation from the sun. The monsoon winds will then reverse and pull drier air masses from the northeast of the regions between October and April (the dry monsoon or the winter monsoon).

In 2022, large sections of Pakistan experienced record-breaking monsoonal rainfall. At one point, as much as one-third of the country was flooded. Many citizens experienced the effects of record flooding from mid-June to the end of August of that year. In August alone, Pakistan received more than three times the usual rainfall. The year 2022 was the wettest year on record since 1961. Sindh and Balochistan, located in the southern regions of Pakistan, recorded rainfall amounts seven to eight times the usual, along with the wettest August in historical records. The record volume of water swelled the Indus River, flooding agricultural-supported rural communities, creating flash floods and high-water levels in urban areas, and saturating the soil, promoting landslides.

According to the National Disaster Management Authority of Pakistan and Voice of America News (the largest US international broadcaster) covering the monsoon of 2022, 33 million people were affected by the extreme rainfall, including a death toll of nearly 1500 people. The urban infrastructure was severely damaged, washing away bridges and roads and destroying homes. In rural areas, crops were damaged beyond salvation, requiring food to be brought in from Iran and other neighboring nations. The floods also created a health crisis. Sherry Rehman, Pakistan's minister of climate change and the International Rescue Committee, reported that Karachi, a southern port city of Pakistan, had an outbreak of dengue fever and other mosquito-borne diseases and sickness symptoms like diarrhea, cholera, skin infections, and malaria (Gul 2022). Overall, the flood damage was estimated at over $30 billion in damages.

Members of the scientific community associated with WWA analyzed the likelihood that the record-breaking torrential monsoon rainfall was heightened by human-caused climate change. Their findings, published in a peer-reviewed article, were that recent trends in the observed records showed that the five-day maximum rainfall over the provinces of Sindh and Balochistan was nearly 75 percent more intense than if the climate had not recently warmed by 1.2 °C. Most models and observations have shown that rainfall in Pakistan has become 50 percent more intense as the climate has warmed. Models project that for climate warming of 2 °C warmer than preindustrial time, rainfall intensity will continue to increase significantly over a five-day event. However, geographical accuracy is difficult to assess for sixty-day monsoon rainfall (Otto et al. 2023).

## Atmospheric Rivers

Atmospheric rivers are a natural part of the hydrologic (water) cycle, bringing precipitation to land masses. They are essential in that they contribute to annual freshwater supplies. They form when large-scale weather patterns align, creating narrow channels of intense moisture transport. They typically start over warm tropical oceans. These narrow air currents, often called "rivers in the sky," transport immense amounts of water vapor from tropical regions to Earth's polar regions and continents. Low-level jet streams push them toward the coast ahead of cold fronts to areas adjacent to the tropics. Atmospheric rivers occur in various regions around the world, including the North Pacific or Atlantic, southeastern Pacific, South Atlantic, and even near Greenland and Antarctica. They affect places like Portugal, Western Europe, Chile, and South Africa. Typically, atmospheric rivers span 250 to 375 miles wide and hold as much water vapor as about twenty-five Mississippi Rivers. Upon landfall, particularly in mountainous regions like the Sierra Nevada and the Andes, atmospheric rivers release much of their accumulated water vapor as snow or rain.

Strong atmospheric rivers can cause disastrous flooding. The exceptionally wet air shoots across the ocean and dumps massive volumes of rain or snow upon landfall. When several atmospheric rivers hit a location within a short time span, they are referred to as AR families. The first heavy downpours saturate the ground. Subsequent storms fall on soil that cannot absorb more water, contributing to runoff. Rivers and streams can also rise to capacity, and warm temperatures may cause snowmelt, further increasing flood.

A team of researchers led by Scripps Institution of Oceanography at the University of California San Diego created a scale to characterize the strength and impacts of atmospheric rivers. The scale was developed by F. Martin Ralph, director of the Center for Western Water and Weather Extremes at Scripps, Jonathan Rutz from the National Weather Service, and several other experts. It ranks atmospheric rivers from 1 to 5 and creates the categories "weak," "moderate," "strong," "extreme," and "exceptional." The scale incorporates the amount of water vapor within an atmospheric river. It uses the time frame of 24 to 48 hours as its standard measurement of longevity.

Suppose an atmospheric river lasts in an area for less than 24 hours. In that case, it is demoted by one category, but if it lingers for more than 48 hours, it is raised one level on the scale. This method is based on research showing that the amount of water vapor is only part of the story, and adding the duration of the atmospheric river over a region adds a better assessment tool to classify its impact. A Cat 1 atmospheric river is considered "primarily beneficial," whereas a Cat 5 is considered "primarily hazardous." Cat 1 to 3 atmospheric rivers produce precipitation comparable to a three-day storm with damages of less than $1-1.5 million, while Cat 4 and 5 atmospheric rivers produce extreme runoff that can cause damages over $1 billion (Monroe 2019).

Scientists from NASA and around the world have been investigating the correlation between global warming and the characteristics of atmospheric rivers in the western United States and Europe. To accomplish this, they rely on two key tools: the IPCC's latest climate projections for the year 2100 and a global algorithm that can detect atmospheric river events and measure their length, width, and water vapor transport. The sophisticated algorithms can be used to model simulations from 1950 to the present. After analysis, scientists found that the models accurately depicted atmospheric rivers in the climate of the programmed period. The algorithm was then used to predict future climate conditions in the latter half of the twenty-first century. This helps scientists compare the frequency and characteristics of atmospheric rivers in the current climate with those of future projections. When the algorithm was tested using lower levels of greenhouse gas emissions, the results showed less extreme atmospheric river characteristics, including their size and the amount of precipitation generated.

On the other hand, when researchers, like those at the Lawrence Berkeley Livermore Lab, simulate a scenario where $CO_2$ emissions continue to rise or where little-to-no mitigation occurs by the end of the twenty-first century, atmospheric rivers shift from being "mostly or primarily beneficial" to "mostly or primarily hazardous." They also found that the increase in precipitation is due to AR families mentioned earlier and not from a single storm. Finally, the Berkeley research team concluded that it is more likely that unmitigated atmospheric greenhouse gas levels will also lead to more intense winter season precipitation totals in the coastal regions of the American Southwest (Bobyock 2024).

Complimentary articles written by atmospheric scientists have concluded similar findings, including a paper published in 2020 in *Nature Reviews* (Payne et al. 2020). They used a theoretical framework to synthesize an understanding of atmospheric river dynamics and thermodynamic responses to anthropogenic warming. The study found that increased atmospheric moisture is highly based on the Clausius–Clapeyron equation, or the physical property that for every increase of 1 °C (1.8 °F), the water-holding capacity of the atmosphere increases by about 7 percent. This will enhance the intensity of atmospheric-related precipitation, especially where the area's geography is conducive to precipitation.

### *Case Study: Atmospheric Rivers Impacting California*

Atmospheric rivers are common in California and have positive and negative impacts. While they can relieve drought conditions, they also cause significant damage to infrastructure and landscapes, with losses amounting to millions or even billions of dollars. Luis Gimeno-Sotelo, an atmospheric scientist at the University of Vigo in Spain, highlights that over 85 percent of flooding events on the western coast of the United States are related to atmospheric rivers. The absence of these moisture highways can also lead to a 90 percent increase in drought occurrences (Gimeno-Sotelo 2022). Studies also show that all major Northern California Russian River watershed floods were associated with Cat 5 atmospheric river events (Ralph and Dettinger 2012). Furthermore, Florsheim and Dettinger (2015) found that 81 percent of California Central Valley levee breaks coincided with atmospheric river events.

Michael Dettinger, a scientist at the US Geological Survey, and his team of researchers from Scripps published a paper in 2011 emphasizing the crucial role of atmospheric rivers in California. The state heavily relies on a small number of wet days per year (typically 5 to 15 days) to accumulate most of its annual precipitation totals. Atmospheric rivers usually fuel the largest storms in California. California records some of the highest three-day storm totals in the country, accounting for 20–50 percent of the state's precipitation and streamflow, comparable to the hurricane belt in the southeastern United States.

In 2022, a group of winter storms killed more than a dozen people in California and put tens of thousands more under evacuation orders and watches. Rain reached 5.5 inches in Northern California and flooded all six lanes of Highway 101 in the city of South San Francisco. On January 8, heavy rains and 70-mile-per-hour winds knocked out power for more than 345,000 people in Sacramento.

A year later, in March of 2023, anyone turning on the news witnessed the massive green blob over the map of the United States hovering over California. The green zone indicated a massive atmospheric river producing an extreme flood risk and winds forecasted to be over 100 mph. Classes at universities like UCLA were canceled, and the population was asked to shelter in place. NOAA's Weather Prediction Center issued an "extreme" risk of flash flooding for the southern Sierra Nevada and the Central California coastal range south of Monterey as their computer model projected an even warmer and wetter forecast than had been previously issued. The "extreme" label is the highest risk NOAA and National Weather Service officials can issue. Only about 4 percent of storms get this distinction.

Several inches of rain brought more widespread flood alerts and mudslide concerns. At the same time, new snowfall totals in the highest elevations reached 100 inches, and ridgetop wind gusts are blowing well over 100 mph. Ski resorts in the Lake Tahoe area clocked wind gusts of 139 mph. Torrential rains triggered an urgent "Flash Flood Emergency" in the central California towns of Springville and Porterville, where residents reported three inches of rain coupled with rapid snow melt. Rising rivers forced the evacuation of communities in coastal Santa Cruz County. At the same time, mudslides and heavy rains covered several roads and highways in central California. Nearly 10,000 residents were given orders to evacuate.

During the three-day storm, thirty-four counties were under a state of emergency. The strong winds and saturated ground led to power outages in the state. According to PowerOutage.us, just over 50,000 customers were without power during the height of the storm. Reservoirs released water to avoid spillovers. At higher elevations, 2 to 4 feet of new snow fell above 7,000 feet, and 6 to 8 feet of snow or more fell above 9,000 feet in the central and southern Sierra. The NWS said several of the Sierra's highest passes, including Carson, Sonora, and Tiaoga, received 100 inches of new snow by the end of that weekend.

Experts compared the outputs of fifteen climate models to assess future atmospheric projections for the West Coast of the United States. Most models projected an increase in year-to-year precipitation variability, especially in California. Models projected a decrease in the overall precipitation for the region. However, dry periods will likely be interrupted by a contribution of atmospheric river-related extreme events, especially for areas of south-facing slopes (Gershunov et al. 2019). Dr. Katerina Gonzales, an EPA Climate

Adaptation Advisor, warns about the increasing intensity of atmospheric rivers due to climate change. She explains that rising amounts of water vapor in the atmosphere leads to more intense atmospheric rivers, which are becoming more formidable and bringing heavier precipitation. Dr. Gonzales emphasizes the need for significant investments in green infrastructure, such as floodplains, parks, and rain gardens, to absorb runoff and mitigate the impact of these storms. She highlights that our current infrastructure was designed for a 20th-century climate and needs to be revised to handle the challenges posed by today's more extreme weather events. As these intense atmospheric rivers are just a preview of what is to come, she stresses the importance of adapting to the climate changes we've already set in motion rather than solely relying on improved forecasting.

## Extreme Drought

As much as we hear about hurricanes and flood events, we also see extreme drought and the plight of the people under those conditions. A drought is when precipitation (rain or snow) is less than the amount of water evaporating from an area, resulting in a water shortage. Scientists monitor water balances by observing precipitation, temperature, surface water flow, aquifers (natural rock underground water storage reservoirs), soil moisture, and snowpack. Analysts compare current conditions to hundreds of years of past moisture trends to determine whether current drought conditions result from natural climate variation or anthropogenic climate change.

The increase in global temperatures promotes droughts in many regions for two main reasons: increased evaporation and less snowpack. Droughts are severe in a warmer world with increased evaporation. People have observed this when boiling water; as the temperature of water increases in the pot, more water evaporates. The same physical properties apply to a warmer planet. As temperatures increase, more water from lakes, rivers, oceans, and soil evaporates. A decrease in snowpack can also trigger regional droughts. According to the Center for Climate and Energy Solutions or C2ES, a leading environmental policy think-tank composed of climate experts and policy consultants, climate change increases the odds that the United States and other areas worldwide will be subject to worsening drought conditions.

Most precipitation begins as snow in the upper atmosphere. If temperatures near the Earth's surface are cold, the precipitation will remain frozen, falling to the ground as snow. A warmer atmosphere near the Earth's surface interferes with this process. For example, in mountainous regions of the world, warmer autumns, winters, and springs create atmospheric conditions where a shift has been observed in the type of precipitation received on the ground. Instead of the precipitation remaining as snow, as it descends through lower and lower elevations of the atmosphere, the frozen particles melt and become rain. The rainwater will then flow to lower elevation areas or evaporate instead of being stored as snowpack. A smaller snowpack means water is not "stored" in a frozen form for months, making it unavailable for extended periods. High-elevation cold-desert regions like Salt Lake City, Utah, are especially concerned about losing snowpack, whose water content flows into reservoirs and groundwater stores to be used during the summer when precipitation is minimal. High-elevation snowpack also supplies the Colorado River and its tributaries, providing water for forty million people.

Just as drinking water is threatened for many populations, water for irrigating crops also becomes scarce. As the soil dries, trees become stressed. The combination of aridity and heat leads many western species to succumb to pest infestations like bark beetles. The stress of drought and heat leads to large forest die-offs. A positive feedback loop sets in. Tree death decreases the amount of $CO_2$ that can be withdrawn from the atmosphere. The atmosphere warms and kills more forests. Dry trees become fuel for fire events, which release $CO_2$ into the atmosphere and warm the planet further.

### *Case Study: Drought in the American Southwest*

In a 2020 study in *Science*, researchers studied the link between human-caused global warming and the twenty-first-century megadrought in the Western United States and northern Mexico. They looked at modeled temperature, relative humidity, and precipitation data between 1901 and 2018 with current conditions and in a counter-climate scenario (without the increase of current greenhouse gases). According to the study, human-caused warming accounted for 46 percent of the drought's severity. Attribution studies to global drought are complex because drought is not uniform worldwide. It tends to be more regional (Williams et al. 2020).

In the 2021 report by the IPCC, scientists were highly confident that every 0.5 °C (0.9 °F) of atmospheric warming will cause several regions, including the American Southwest, to experience a rise in the intensity and frequency of droughts. The regions will also see a decline in the health of agricultural landscapes and ecosystems regionwide. The report also concluded that extreme drought events once a decade are now almost twice as likely (1.7 times) than they were from 1850 to 1900 before humans impacted the climate.

### *Case Study: Southern African Drought, 2018–2021*

The geographic location of the southern portion of Africa lends it to a variable climate, including times of drought. The wet season is brought on by strong rainfall patterns expected in the region from November to March. During the wet season, high amounts of precipitation infiltrate and saturate the soils. During a robust wet season, soils can retain moisture long into the dry months. Natural variations in wind and ocean patterns like those of El Niño events (warming of sea surface temperatures in the central and eastern Equatorial Pacific) can disrupt the region's rainfall amounts during the crucial months when water stores are replenished. Ongoing El Niño conditions created severe drought-stricken areas in Southern Africa beginning in 2012. In Namibia, nearly 42 percent of people faced food insecurity during the drought of 2012/2013, prompting government officials to declare a national state of emergency. Southern Africa faced an extreme drought beginning in October of 2018. Nations on the southern tip of Africa rely on an economy of climate-sensitive agriculture.

As drought conditions continued, organizations such as the Southern African Development Community and the Food and Agriculture Organization (FAO 2018) advised farmers to prepare for further drought conditions and stockpile their goods; however, the warning did not avoid the severe food insecurity faced by 9.6 million people in 2019 (FAO 2018). By August 2019, the drought had been classified as a level 2 Red-Class event by the Global Disaster Alert and Coordination System. The drought subsided briefly when Southern Africa experienced above-average rainfall in early 2022 (South African Weather Service 2023). The United Nations Development Program advises developing countries on the cumulative risks associated with climate change. They warn that higher temperatures and a reduction in rainfall expected as a result of climate change will continue to reduce the availability of water, especially during El Niño years.

## Increased Wildfires

Wildfires conjure up images of a vast swath of forest consumed in flames. News channels and online pop-ups show the apocalyptic scene, especially during the late summer months after a dry year or a prolonged drought. Global warming is increasing fire activity around the world. Although forest fires are started by various ignition sources, some natural and some anthropogenic, they are made more likely, intense, and frequent under hot, dry conditions. These reasons are valid in Australia and the United States. In other regions and countries such as Africa, Central and South America, México, Fiji, New Zealand, and Southeast Asia, humans are mainly responsible for forest fires through agricultural and commercial practices like farming and slash-and-burn agriculture. On a smaller scale, human activity can be a driver of fire or fire suppression. Still, scientists have run models projecting future wildfire scenarios.

The northern regions of our planet have already seen unprecedented fire activity in the past few years. Boreal forests, also called taiga or snow forests, are biomes characterized by coniferous forests with pines, spruces, and larches. As the high latitudes of the Northern Hemisphere (areas above 60 ° latitude) warm at three times the rate of the rest of the planet, the boreal forests are becoming more arid. With global warming, the areas with frequent fire-prone conditions will increase by 29 percent, mainly in the boreal (+111 percent) and temperate (+25 percent) zones, potentially lengthening the fire season. Their estimates of the expansion of fire-prone regions are uneven and more pronounced in the northern regions by a large percentage. With added warmth, boreal forests may spread over a larger area, reaching most of Northern Scandinavia and undergoing a southward and northward expansion in Canada, Alaska, and Russia. The expansion would alter the potential, distribution, and severity of wildfires.

Models also show that temperature rather than precipitation will predominantly promote fire events as the year 2100 approaches (Senande-Rivera et al. 2022). The most remarkable changes in the temperate regions are expected in Southern China and Southern Europe. Much of Europe is projected to transition from an infrequent fire category to a more frequent one.

### *Ecological Damage from Wildfires*

Several studies by fire experts have examined the effects of climate change and wildfires on the composition of forest ecosystems. Wildfires can change vegetation by killing adult trees established before the current regional temperatures. Seedlings, however, have been noted to have difficulty establishing themselves and surviving in a warmer climate. Researchers have pointed out declines in postfire numbers of particular species of trees. Scientists examined the relationship between annual climate conditions and postfire tree regeneration for regions with conifers common in low-elevation areas of the Southwest (ponderosa pine and Douglas fir).

A study by Davis et al. (2024) of thirty-three wildfire areas across four regions in the western United States revealed a direct response to climate conditions, including soil moisture and maximum surface temperature. Drier conditions over the past twenty years have made it more difficult for seedlings to survive. The researchers note that conditions have become increasingly unsuitable for regeneration, including high fire severity (a fire producing a high percentage of tree death— >80 percent—and extensive mineral soil exposure) and low seed availability. They then conclude that climate change combined with high-severity fire diminishes opportunities for seedlings and may lead to an overall shift in vegetation in low-elevation ponderosa pine and Douglas fir forests across the western United States.

Wildfires also disturb animal communities. For example, the Australian bushfires of 2019 killed or displaced an estimated three billion animals. The Worldwide Fund for Nature called it the worst wildlife disaster in modern history. Mammals, reptiles, birds, and frogs died in the fires. A recent report estimated that koalas may face extinction in New South Wales by 2050.

### *Case Study: Canadian Wildfires Blanket the United States with Smoke*

Scorching temperatures in many parts of Canada have contributed to increased wildfires. As temperatures rise, leaves, grass, twigs, and downed logs get drier and drier and become potential fuel. Dry fuel burns longer and hotter. No matter the cause of the ignition, climate change makes wildfires more extensive and more severe because the plants that become fuel for fires are drier in an area under increased heat and drought conditions.

In 2019, the Canadian National Assessment Report and the Center for Climate and Energy Solutions looked at the impact of climate change on the country, including the uptick in wildfires. The studies concluded that the dry organic fuel matter generated from unseasonably dry and hot weather helped promote more extensive wildfires than usual. The Natural Resources Agency of

Canada reported that Canadian fires have been four times larger and three times more frequent since 2000.

Canadian wildfires also characterized the year 2021. When interviewed, Ellen Whitman, forest fire research scientist with Natural Resources Canada, reported that the nationwide area burned was more than 1.5 times the average in recent history. The cause, according to the fire experts, is an early warm spring. Warmer and drier conditions with minimal snowpack and little winter rains desiccate vegetation and fuel more fire. When burned, the drier fuels create hotter, more extensive, longer-lasting fires.

In 2023, over 75 million Americans were affected by wildfires from Canada. That year, Canada experienced its warmest May to June period since 1940, and at a national scale, relative humidity was very low. The warm and dry conditions and continuous southeasterly winds fueled extensive fires in Alberta, British Columbia, central Saskatchewan, and southwestern portions of the Northwest Territories. During that spring and summer, the extensive wildfires burned over thirteen million hectares in Canada. The fires caused seventeen fatalities and the displacement of more than 150,000 people. Hundreds of structures were damaged. In southern Ontario, air quality reached the "very high risk" level, forcing public event cancellations. Nova Scotia, New York, New Jersey, and Connecticut suspended schools for several days.

Smoke from wildfires in Canada drifted into the United States, leading to inferior air quality in the Midwest and across much of the eastern states, from New England to the Southeast. According to the Environmental Protection Agency, more than 100 million Americans were given air quality warnings. IQAir, an air quality monitoring website, reported New York City's pollution as one of the worst in the world, fourth behind New Delhi, India; Dhaka, Bangladesh; and Toronto with a 484 out of 500 rating on the Air Quality Index scale.

## Extreme Heatwaves

According to the NOAA, a heat wave is unusually hot weather typically lasting two or more days. The temperatures must be outside the historical averages for the area. This is similar to the idea of "record-breaking" temperatures. In 2012, the United States experienced a heat wave. Over 8,000 warm temperature records were broken or tied. Chicago roads buckled, and many deaths were attributed to the nationwide heat.

Heat waves generally form when air masses are trapped over a region. For example, during the 2012 heat wave, the air was stagnant and did not move for days. A high-pressure system from Mexico had moved north. It then settled over the Great Plains of the United States and stalled. Instead of circulating the globe, the air hovered over the same states and cities, warming like the air inside an oven. High-pressure systems force air downward, not letting the air close to the ground rise, trapping the warm ground air. Since the air could not rise, no rain could form, and the air got hotter and hotter.

Heat extremes have become more frequent and more intense. Heat extremes will continue to increase in all continents with the continued addition of greenhouse gases even if global warming does not exceed an increase of 1.5 °C from preindustrial levels. If temperatures increase by 2 °C, extreme heat events will likely double in intensity. At an increase of 3 °C, extreme events will quadruple (IPCC 2021, 1518). Regional differences around the world amplify the human-enhanced greenhouse effect. For example, in some regions, the loss of ice or snow cover exposes land surfaces that are not as efficient at reflecting incoming solar radiation (albedo), raising temperatures in those regions.

### *Case Study: Extreme Heat Waves in Europe*

As stated earlier, the heat waves in Europe triggered the first attribution study (Stott, Stone, and Allen 2004). Those heat waves killed 70,000 Europeans in one month, with the highest rates of mortality in Italy, Greece, Spain, and Portugal. The authors concluded that it is very likely (confidence level > 90 percent) that human influence at least doubled the risk of that heatwave.

A severe heatwave occurred from June to July 2019 over western-central Europe, the most densely populated region. Attribution studies determined that anthropogenic climate change caused a sevenfold increase in the chance of a particular event. They also surmised that those events will become very common, with the return periods of 1.8–7.2 years (Wang et al. 2020). Sure enough, it was not long until the next record-breaking heatwave. In the summer of 2022, 61,672 people died. The mortality was 56 percent higher among women in comparison to men.

Yet again, 2023 started off hot in Europe. On day 1 (January 1, 2023), more than eight European countries broke heat records for the warmest January day ever, including Liechtenstein, the Czech Republic, Poland, the Netherlands, Belarus, Lithuania, Denmark, and Latvia. Scientists attributed those heat waves to climate change. Later that year, Albania set its all-time high-temperature record at 44 °C (111 °F) in Kuçovë. Cyprus neared 45 °C (113 °F) during the heatwave. Toulouse, France, reached 42.4 °C (108.3 °F). The World Meteorological Organization station in Thiva registered 44.2 °C (111.6 °F), making it the highest temperature in Athens, recorded at 43.4 °C (110.1 °F). Officials announced the midday closure of the Acropolis. The National Observatory of Athens station in Elafonisi recorded a maximum temperature of 44.2 °C (111.6 °F). Temperatures continued to soar throughout the year. The UK broke record after record in September, with health officials issuing warnings and closures. Government officials issued heat warnings and set up "heat shelters" in various cities. All throughout Europe, temperatures rose to new levels.

According to Climate Central, more than 80 percent of people on Earth experienced a sweltering July in 2023, and the extreme heat was attributed to climate change (Climate Central 2023). In fact, each day, two billion people experienced heat waves made three times more likely because of climate change. Published papers reported that 6.5 billion people experienced extreme heat at least one day that July. Temperatures were extremely high in the southern United States, Southern Europe, and Southeast Asia. NASA senior scientist Katherine Calvin confirmed what atmospheric scientists have long known: the likelihood of such events in the future depends on future emissions.

### *Case Study: Extreme Heatwaves in Africa*

One of the most urgent topics at the IPCC's annual meeting in November of 2022 in Egypt (COP27) was Africa and extreme weather events. The extreme weather in Africa is mainly unreported in American or European broadcasts. However, less visible reports by organizations like NOAA reported that 2021 tied 2019 as Africa's third warmest on record (Vose et al. 2021). Carbon Brief, an organization that analyzes disaster records and models climate events, found that extreme weather events in Africa have been prevalent and, at times, unprecedented (Tandon 2023). Since the start of 2022, at least 4,000 African people have lost their lives, while over nineteen million were affected by heatwaves, flooding, and droughts. Uganda alone lost 2,500 people to drought and famine. In Ethiopia, eight million people were affected by drought-induced famine. On the opposite spectrum of drought is the country of Nigeria, which experienced floods that killed more than 600 people. Chad suffered torrential rains and floods in August and October. Madagascar and Mozambique have been in the path of severe storms, with nearly 900 people killed in 2022. Heat waves in Tunisia reached 48 °C in July, spawning wildfires throughout the country (NASA 2022).

## Extreme Cold Spells

Can places still experience record-low temperatures while the planet continues to warm? It makes sense why a hotter planet leads to record high temperatures, wildfires, and droughts, but how do scientists explain a cold winter in a world that scientists also confirm is getting hotter? While high temperatures occur more frequently than cold extremes (Washington et al. 2009), anomalous temperatures are recorded in several regions. Place after place has been breaking records during the winter due to blizzard conditions and sustained cold spells. In the early winter of 2024, cold, arctic air descended on large portions of the United States and across the United States to the southern regions, where extreme winter weather is rare. Texans were asked to conserve energy to avoid electrical outages, and people in Arkansas experienced record-low temperatures. These snowy conditions and record-cold temperatures followed the hottest year on record.

The term "global warming" confuses many, especially when the overall warming of the planet and the ocean can lead to extreme cold events with high precipitation. When scientists and the media first used the term "global warming" a few decades ago, they specifically referred to the process of greenhouse gases trapping heat in the atmosphere and warming the planet's average temperature. However, record-high temperatures in many places have just one impact on this decades-long shift. With climate modeling, scientists now understand that an atmosphere with rising levels of carbon and methane leads to more disastrous weather on both the warm and cold extremes as the effects of climate change play out.

Climate change is a better term for encompassing the shifts in continuous weather events that are incredibly different from what a location historically experiences during a particular time of year, including winter. However, when a severe winter storm blankets the ground, many mistakenly reply, "Bring back global warming; it is freezing out here!" However, scientists have studied these occurrences and are securing evidence that global warming has led to extreme amounts of snow and temperatures that have sunk to minimum degrees for specific locations during the winter. Despite this, it is essential to remember that global warming has led to an alarming trend of disappearing snow in the Northern Hemisphere over the long term.

### *Changes to the Northern Jetstream*

How is global warming changing the winter climate of places like the American Midwest, the American East Coast, and northern European cities with the addition of more frequent and dangerous cold weather and even storms nicknamed bomb cyclones? What is the relationship or link between the two? The first relationship is that of the condition of the Arctic areas of the world to the jet stream position and winter weather. The Arctic is warming three to four times faster than the rest of the world.

In the winter of 2021, the Arctic Sea ice extent was at a record low for winter levels, and temperatures in the coldest months were sixty degrees above normal. As the Arctic warms, the temperature difference between the polar regions and the high and mid-latitudes decreases. The closer the temperatures are in the Arctic to those in the continental United States or Europe, the more the jet stream (a highway-like system of strong winds 6 miles (10 km) above Earth's surface) weakens in its speed and circular path around the North Pole. It will then slow down and reach or kink southward, bringing Arctic air masses (air with the same characteristics in temperature and humidity) into the United States.

High in the stratosphere spins another highway of cold air that meteorologists and climate scientists refer to as a polar vortex. This belt of strong winds spins around the North Pole like a cap on the Northern Hemisphere. The polar vortex is sometimes described as a spinning top. In its average state, it rotates very fast, keeping the frigid northern air locked in the Arctic region. However, it can get disrupted and knocked off course, becoming stretched and distorted, spilling cold air and influencing the jet stream's path.

This idea of a weakened jet stream pulling cold northern air to the mid-latitudes and beyond was initially proposed by Jennifer Francis of the Woodwell Climate Research Center in Massachusetts. The idea has been gaining traction since 2012. Her research highlighted that the closer temperatures in the Arctic are to the south, the wavier the Jetstream becomes, pushing frigid air southward (Devlin 2015). Research continues on this area of study, including the 2021 study by Judah Cohen, a climatologist at the Massachusetts Institute of Technology. He also found that the rapid heating in parts of the Arctic, combined with high snowfall in Siberia, made the jet stream wavier and knocked the polar vortex off course ("Study Reveals" 2021).

While there have been several freezing Northern Hemisphere winters coinciding with warm Arctic winters, it is difficult to determine cause and effect, according to James Screen, professor of climate science at Exeter University. Screen's research concluded the opposite. He argues that over the long term, global warming is leading to fewer and less severe cold extremes (Paddison, Shveda, and Razo 2023).

The second link to global warming is the relationship of a warmer atmosphere to how much moisture it can hold. As we discussed earlier in the section on hurricanes, warm air can hold more moisture. As the atmosphere warms, it holds but also releases more moisture faster than it did several decades ago. A warmer ocean adds to this condition. Just as increased evaporation adds more water vapor to the air in the summertime, fueling hurricanes with added moisture and increasing rainfall, they also add to winter cyclonic activity, increasing snowfall rates and amounts during the winter (especially near the coast).

### *Case Study: Bomb Cyclones from Snowpocalypse to Snowzilla*

Dangerous blizzard conditions blasted the Northeast Coast of the United States in January 2022 with heavy snowfall totals of 1 to 2 feet from the mid-Atlantic states to Canada. The blizzards cost millions of dollars in travel revenue and knocked out power to over 100,000 people. Jonathan Overpeck, a climate research scientist at the University of Michigan, reminded the public of similar storms nicknamed Snowpocalypse, Snowmageddon, and Snowzilla, which have hit the eastern United States over the past ten years. He explains that these storms, often called "bomb cyclones" because of the swift drop in atmospheric pressure and quick development, are made more substantial and more frequent by global warming (Overpeck 2022).

## Regions Susceptible to Climate Change and Extreme Weather Events

Several regions of the world are especially vulnerable to extreme weather events caused by climate change. The IPCC has identified Africa, South Asia, the Pacific Islands, and the Caribbean as vulnerable populations of the world where conditions are challenging even before extreme events amplify the struggle for basic needs such as food, shelter, and access to healthcare. Droughts, heat waves, storms, and flooding are projected to amplify the socioeconomic inequalities. More extreme weather means communities struggling to make ends meet will continue to struggle and be worse than before a weather event. The lack of governmental support in developing nations whose cities or rural populations do not have access to relief operations makes it very difficult for people in these nations to recuperate or rebuild (IPCC 2023). Chapter 2 digs deeper into this issue and discusses the ethical responsibility of high-emitting countries to the least-emitting countries that are most vulnerable.

## Further Reading

Alexander, R. 2023. "Mainstream Gets Extreme Weather Wrong Again." Clintel, February 3, 2023. Accessed October 14, 2024. https://clintel.org/mainstream-gets-extreme-weather-wrong-again.

American Meteorological Society. n.d. "Explaining Extreme Events from a Climate Perspective." Accessed April 12, 2023. https://www.ametsoc.org/index.cfm/ams/publications/bulletin-of-the-american-meteorological-society-bams/explaining-extreme-events-from-a-climate-perspective.

"Attribution Science: Linking Climate Change to Extreme Weather." 2021. Columbia Climate School, April 27, 2021. Accessed October 14, 2024. https://oregonclimateeducation.org/resource/attribution-science-linking-climate-change-to-extreme-weather.

Bobyock, J. "Unraveling the Mysteries of Consecutive Atmospheric River Events." Phys.org, April 18, 2024. Accessed October 14, 2024. https://phys.org/news/2024-04-unraveling-mysteries-consecutive-atmospheric-river.html.

Calvin, K. "Heatwaves in Europe: A Summer of Record Temperatures and Mortality." *Nature* 398, no. 3 (2023): 123–40.

Cho, R. "Attribution Science: Linking Climate Change to Extreme Weather." Columbia Climate School, Climate, Earth, and Society, October 4, 2021. Accessed October 4, 2024. https://news.climate.columbia.edu/2021/10/04/attribution-science-linking-climate-change-to-extreme-weather.

Climate Central. 2023. "Worldwide Daily Fingerprints of Climate Change during Earth's Hottest Month." Climate Central Report, August 1, 2023.

Climate Central. n.d. "Rising Seas Threaten Coastal Cities." Accessed July 31, 2024. https://www.climatecentral.org.

Colbert, A. 2022. "A Force of Nature: Hurricanes in a Changing Climate." NASA, June 1, 2022. Accessed October 14, 2024. https://climate.nasa.gov/news/3184/a-force-of-nature-hurricanes-in-a-changing-climate.

Davis, K. T., J. Peeler, J. Fargione, R. D. Haugo, K. L. Metlen, M. D. Robles, and T. Woolley. 2024. "Tamm Review: A Meta-analysis of Thinning, Prescribed Fire, and Wildfire Effects on Subsequent Wildfire Severity in Conifer-dominated Forests of the Western US." *Forest Ecology and Management* 561: 121885.

Devlin, D. 2015. “Rutgers Climate Scientists Find More Evidence Linking Arctic Warming to Jet Stream Movement.” Rutgers University, June 1, 2015. Accessed October 14, 2024. https://www.rutgers.edu/news/rutgers-climate-scientists-find-more-evidence-linking-arctic-warming-jet-stream-movement.

Emanuel, K. 2017. “Assessing the Present and Future Probability of Hurricane Harvey’s Rainfall.” *Proceedings of the National Academy of Sciences* 114, no. 48 (November 13): 12681–4. https://doi.org/10.1073/pnas.1716222114.

Eyring, V., N. P. Gillett, K. M. Achuta Rao, R. Barimalala, M. Barreiro Parrillo, N. Bellouin, C. Cassou, P. J. Durack, Y. Kosaka, S. McGregor, S. Min, O. Morgenstern, and Y. Sun. 2021. “Human Influence on the Climate System.” In *Climate Change 2021: The Physical Science Basis, Contribution of Working Group I to the Sixth Assessment Report of the Intergovernmental Panel on Climate Change*, 423–552. Cambridge, UK: Cambridge University Press. https://doi.org/10.1017/9781009157896.005.

“FAO Warns Southern Africa to Prepare for Drought.” 2018. Voices of Africa, September 5, 2018. Accessed 14 October 2024. https://www.voanews.com/a/fao-warns-southern-africa-to-prepare-for-drought/4559235.html.

Fletcher, C. 2019. *Climate Change: What the Science Tells Us.* New York: Wiley.

Florsheim, J. L., and M. D. Dettinger. 2015. “Promoting Atmospheric-River and Snowmelt-Fueled Biogeomorphic Processes by Restoring River-Floodplain Connectivity in California’s Central Valley.” In *Geomorphic Approaches to Integrated Floodplain Management of Lowland Fluvial Systems in North America and Europe*, edited by P. F. Hudson and H. Middelkoop, 119–36. New York: Springer.

Gershunov, A., T. Shulgina, R. E. Clemesha, D. R. Cayan, S. D. Polade, J. Kalansky, and F. M. Ralph. 2019. “Precipitation Regime Change in Western North America: The Role of Atmospheric Rivers.” *Scientific Reports* 9, no. 1 (July 9). https://doi.org/10.1038/s41598-019-46169-w.

Ghali, R., and M. A. Akhloufi. 2023. “Deep Learning Approaches for Wildland Fires Using Satellite Remote Sensing Data: Detection, Mapping, and Prediction.” *Fire* 6, no. 5: 192. Accessed 14 October 2024. https://www.frames.gov/catalog/68203.

Gibbens, S. 2024. “Why Cold Weather Doesn’t Mean Climate Change Is Fake.” *National Geographic*, January 16, 2024. Accessed October 14, 2024. https://www.nationalgeographic.com/environment/article/climate-change-colder-winters-global-warming-polar-vortex.

Gimeno-Sotelo, L. 2022. “Concurrent Extreme Events of Atmospheric Moisture Transport and Continental Precipitation: The Role of Landfalling Atmospheric Rivers.” *Atmospheric Research* 278 (November), 106356.

“Global Warming Could Cause Colder Winters.” 2012. *Physics Today*, January 13, 2012. https://doi.org/10.1063/pt.5.025818.

Gonzales, Katerina, and Daniel Swain. “Episode 45: Atmospheric Rivers.” *Water Talk*, produced by Water Talk, 2023. Podcast, MP3 audio, 00:00:00. https://www.water-talk.com/episode45.

Gul, A. 2022. "Pakistan Blames Climate Change for Deadly Floods, Declares Emergency." Voice of America, August 26, 2022. Accessed October 14, 2024. https://www.voanews.com/a/pakistan-blames-climate-change-for-deadly-floods-declares-emergency-/6718394.html.

"Half of the Country Is under Water; Pakistan Sought International Help." 2022. Kerala9.com, August 29, 2022. Accessed October 24, 2024. https://www.kerala9.com/latest-news/world-news/half-of-the-country-is-under-water-pakistan-sought-international-help.

Halverson, J. B. 2018. "The Costliest Hurricane Season in US History." *Weatherwise* 71, no. 2 (February 15): 20–7. https://doi.org/10.1080/00431672.2018.1416862.

Hernández-Delgado, E. A., P. Alejandro-Camis, G. Cabrera-Beauchamp, J. S. Fonseca-Miranda, N. X. Gómez-Andújar, P. Gómez, R. Guzmán-Rodríguez, I. Olivo-Maldonado, and S. E. Suleimán-Ramos. 2024. "Stronger Hurricanes and Climate Change in the Caribbean Sea: Threats to the Sustainability of Endangered Coral Species." *Sustainability* 16, no. 4 (February 9): 1506. https://doi.org/10.3390/su16041506.

Hibbert, C. McCormick. 2023. "California Is Once Again Being Deluged by Atmospheric Rivers. What Are They, and Will Climate Change Make Them Worse?" *Scientific American*, March 20, 2023.

Hounkpè, J., B. Diekkrüger, D. Badou, and A. Afouda. 2016."Change in Heavy Rainfall Characteristics over the Ouémé River Basin, Benin Republic, West Africa." *Climate* 4, no. 1 (March 4): 15. https://doi.org/10.3390/cli4010015.

Intergovernmental Panel on Climate Change (IPCC). 2014. *Climate Change 2014: Synthesis Report. Contribution of Working Groups I, II and III to the Fifth Assessment Report of the Intergovernmental Panel on Climate Change*. Geneva, Switzerland.

Intergovernmental Panel on Climate Change (IPCC). 2021. *Climate Change 2021: The Physical Science Basis. Contribution of Working Group I to the IPCC Sixth Assessment Report*. Cambridge, UK: Cambridge University Press.

Intergovernmental Panel on Climate Change (IPCC). 2023. *Climate Change 2022—Impacts, Adaptation and Vulnerability*. June 22, 2023. https://doi.org/10.1017/9781009325844.

Jones, N. 2023. "The Ocean Is Hotter Than Ever: What Happens Next?" *Nature* 617, no. 7961 (May 10): 450. https://doi.org/10.1038/d41586-023-01573-1.

Li, S., and F. E. L. Otto. 2022. "The Role of Human-induced Climate Change in Heavy Rainfall Events Such as the one Associated with Typhoon Hagibis." *Climatic Change* 172: 7. https://doi.org/10.1007/s10584-022-03344-9.

Mann, M. E., and K. A. Emanuel. 2006. "Atlantic Hurricane Trends Linked to Climate Change." *Eos, Transactions American Geophysical Union* 87, no. 24 (June 13): 233–41. https://doi.org/10.1029/2006eo240001.

Monroe, R. 2019. "Atmospheric River Storms Create $1 Billion-a-Year Flood Damage." Scripps Institution of Oceanography, December 4, 2019. Accessed October 14, 2024.

https://scripps.ucsd.edu/news/atmospheric-river-storms-create-1-billion-year-flood-damage.

NASA. 2022. "Heatwaves and Fires Scorch Europe, Africa, and Asia." July 13, 2022. Accessed October 14, 2024. https://earthobservatory.nasa.gov/images/150083/heatwaves-and-fires-scorch-europe-africa-and-asia.

NASA. n.d. "Global Climate Change: Evidence." NASA Global Climate Change and Global Warming: Vital Signs of the Planet. Accessed January 14, 2015. http://climate.nasa.gov/evidence.

Nisbet, E. G. 2023. "Climate Feedback on Methane from Wetlands." *Nature Climate Change* 13, no. 5 (March 20): 421–2. https://doi.org/10.1038/s41558-023-01634-3.

Nisbet, E. G., R. E. Fisher, D. Lowry, J. L. France, G. Allen, J. Pyle, M. E. Smith, R. A. Mackenzie, S. Mihalopoulos, R. M. Holmes, F. P. N. Dasari, M. E. Douglas, S. J. Ewings, S. O'Doherty, C. E. Grant, P. B. Butler, C. A. Worrall, A. J. Wright, and D. E. Shallcross. 2016. "Rising Atmospheric Methane: 2007–2014 Growth and Isotopic Shift." *Global Biogeochemical Cycles* 30 (9): 1246–60. https://doi.org/10.1002/2016GB005406.

NOAA National Centers for Environmental Information (NCEI). 2024. "U.S. Billion-Dollar Weather and Climate Disasters." Accessed October 14, 2024. https://www.ncei.noaa.gov/access/billions/, https://doi.org/doi10.25921/stkw-7w73.

Oldenborgh, G. J. van, S. Philip, E. Aalbers, R. Vautard, F. Otto, K. Haustein, F. Habets, R. Singh, and H. Cullen. 2016. "Rapid Attribution of the May/June 2016 Flood-inducing Precipitation in France and Germany to Climate Change." June 22, 2016. https://doi.org/10.5194/hess-2016-308.

Otto, F. E. L., M. Zachariah, F. Saeed, A. Siddiqi, S. Kamil, H. Mushtaq, T. Arulalan, K. AchutaRao, S. T. Chaithra, C. Barnes, S. Philip, S. Kew, R. Vautard, G. Koren, I. Pinto, P. Wolskill, M. Vahlberg, R. Singh, J. Arrighi, M. van Aalst, L. Thalheimer, E. Raju, S. Li, W. Yang, L. J. Harrington, and B. Clarke. 2023. "Climate Change Increased Extreme Monsoon Rainfall, Flooding Highly Vulnerable Communities in Pakistan." *Environmental Research: Climate* 2, no. 2. https://doi.org/10.1088/2752-5295/acbfd5.

Overpeck, J. 2022. "More 'Snowmageddon' and 'Bomb Cyclone' Winter Storms Are in Our Future." University of Michigan, February 1, 2022. Accessed October 14, 2024. https://seas.umich.edu/news/more-snowmageddon-and-bomb-cyclone-winter-storms-are-our-future.

Paddison, L., K. Shveda, and J. Razo. 2023. "Extreme Cold Snaps: Why Temperatures Still Plummet to Dangerous Levels Even as the Planet Warms." CNN, February 3, 2023. Accessed October 14, 2024. https://kesq.com/news/national-world/cnn-world/2023/02/03/extreme-cold-snaps-why-temperatures-still-plummet-to-dangerous-levels-even-as-the-planet-warms-2.

Park, G. 2021. "A Comprehensive Analysis of Hurricane Damage across the U.S. Gulf and Atlantic Coasts Using Geospatial Big Data." *ISPRS International Journal of Geo-Information* 10, no. 11 (November 17): 781. https://doi.org/10.3390/ijgi10110781.

Pasch, R. J., A. B. Penny, and R. Berg. 2018. "Tropical Cyclone Report: Hurricane Maria." Report, AL152017, April 5, 2018. Miami, FL: National Hurricane Center.

Patricola, C. M., and M. F. Wehner. 2018. "Anthropogenic Influences on Major Tropical Cyclone Events." *Nature* 563, no. 7731: 33946.

Payne, A. E., M.-E. Demory, L. R. Leung, A. M. Ramos, C. A. Shields, J. J. Rutz, N. Siler, G. Villarini, A. Hall, and F. M. Ralph. 2020. "Responses and Impacts of Atmospheric Rivers to Climate Change." *Nature Reviews Earth & Environment* 1, no. 3 (March 9): 143–57. https://doi.org/10.1038/s43017-020-0030-5.

Ralph, F. M., and M. D. Dettinger. 2012. "Future Changes in the Hydrologic Cycle Associated with Flood-Producing Storms in California." *Journal of Hydrometeorology* 13, no. 2: 706–17.

Ramos-Scharrón, Carlos E., and E. Arima. 2019. "Hurricane María's Precipitation Signature in Puerto Rico: A Conceivable Presage of Rains to Come." *Scientific Reports* 9, no. 1 (October 30). https://doi.org/10.1038/s41598-019-52198-2.

Reed, K. A., M. F. Wehner, and C. M. Zarzycki. 2022. "Attribution of 2020 Hurricane Season Extreme Rainfall to Human-Induced Climate Change." *Nature Communications* 13, no. 1 (April 12). https://doi.org/10.1038/s41467-022-29379-1.

Risser, M. D., and M. F. Wehner. 2017. "Attributable Human-induced Changes in the Likelihood and Magnitude of the Observed Extreme Precipitation during Hurricane Harvey." *Geophysical Research Letters* 44, no. 24 (December 23). https://doi.org/10.1002/2017gl075888.

Schaller, N., A. Kay, R. Lamb, N. R. Massey, G. J. van Oldenborgh, F. E. L. Otto, S. N. Sparrow, R. Vautard, P. Yiou, I. Ashpole, A. Bowery, S. M. Crooks, K. Haustein, C. Huntingford, W. J. Ingram, R. G. Jones, T. Legg, J. Miller, J. Skeggs, D. Wallom, A. Weisheimer, S. Wilson, P. A. Stott, and M. R. Allen. 2016. "Human Influence on Climate in the 2014 Southern England Winter Floods and Their Impacts." *Nature Climate Change* 6: 627–34. https://doi.org/10.1038/nclimate2927.

Senande-Rivera, M, D. Insua-Costa, and G. Miguez-Macho. 2022. "Spatial and Temporal Expansion of Global Wildland Fire Activity in Response to Climate Change." *Nature News*, March 8, 2022. Accessed October 14, 2024. https://www.nature.com/articles/s41467-022-28835-2.

South African Weather Service. 2023. *Annual State of the Climate of South Africa 2022.* Pretoria. South Africa.

Stott, P. A., D. A. Stone, and M. R. Allen. 2004. "Human Contribution to the European Heatwave of 2003." *Nature* 432, no. 7017 (December 2): 610–14.

"Study Reveals Extreme Winter Weather Is Related to Arctic Change." 2021. Phys.org, September 2, 2021. Accessed October 14, 2024. https://phys.org/news/2021-09-reveals-extreme-winter-weather-arctic.html.

Tandon, A. 2023. "Eastern Canada Wildfires: Climate Change Doubled Likelihood of 'Extreme Fire Weather.'" Carbon Brief, September 14, 2023. Accessed October 14, 2024. https://www.carbonbrief.org/eastern-canada-wildfires-climate-change-doubled-likelihood-of-extreme-fire-weather.

Taylor, M. A., L. A. Clarke, A. Centella, A. Bezanilla, T. S. Stephenson, J. J. Jones, J. D. Campbell, A. Vichot, and J. Charlery. 2018. "Future Caribbean Climates in a World of Rising Temperatures: The 1.5 vs. 2.0 Dilemma." *Journal of Climate* 31, no. 7 (April): 2907–26. https://doi.org/10.1175/jcli-d-17-0074.1.

Trenberth, K. E., L. Cheng, P. Jacobs, Y. Zhang, and J. Fasullo. 2018. "Hurricane Harvey Links to Ocean Heat Content and Climate Change Adaptation." *Earth's Future* 6, no. 5 (May): 730–44. https://doi.org/10.1029/2018ef000825.

US Environmental Protection Agency. 2020. "Understanding Global Warming Potentials." EPA, September 9, 2020. Accessed October 14, 2024. https://19january2021snapshot.epa.gov/ghgemissions/understanding-global-warming-potentials_.html.

US Environmental Protection Agency. 2024. "Climate Change Indicators in the United States." Fifth edition, EPA 430-R-24-003.

Vose, R., B. Huang, X. Yin, D. Arndt, D. R. Easterling, J. H. Lawrimore, M. J. Menne, A. Sanchez-Lugo, and H. M. Zhang. 2021. Implementing Full Spatial Coverage in NOAA's Global Temperature Analysis. Geophysical Research Letters 48(10), e2020GL090873; https://doi.org/10.1029/2020gl090873.

Wang, J., J. Feng, Z. Yan, and Y. Chen. 2020. "Future Risks of Unprecedented Compound Heat Waves over Three Vast Urban Agglomerations in China." *Earth's Future* 8: e2020EF001716.

Washington, W. M., Reto Knutti, Gerald A. Meehl, Haiyan Teng, Claudia Tebaldi, David Lawrence, Lawrence Buja, and Warren G. Strand. 2009. "How Much Climate Change Can Be Avoided by Mitigation?" *Geophysical Research Letters* 36, no. 8: L08703.

Whitman, E. 2023. "Impact of Wildfires on Canadian Forests." *Canadian Fire Science Quarterly* 23, no. 2: 45–58.

Williams, P. A., B. I. Cook, J. E. Smerdon, R. Seager, J. T. Abatzoglou, and E. R. Cook. 2020. "Large Contribution from Anthropogenic Warming to an Emerging North American Megadrought." *Science* 368, no. 6488: 314–18.

World Meteorological Organization (WMO). n.d. "Extreme Weather." Accessed July 22, 2024. https://wmo.int/topics/extreme-weather.

# 2

# Problems, Controversies, and Solutions

In Chapter 1 we discussed the warming of the planet due to a surplus of greenhouse gases by roughly 1 °C (2 °F) since the pre-industrial era (1850–1900) and how extreme weather events can now be statistically attributed to an increase in global surface temperatures of land and the ocean. Given the tremendous size and heat capacity of the Earth's oceans, it takes a massive amount of added heat energy to raise Earth's average yearly surface temperature even a fraction of a degree. The increase in global average surface temperature that has occurred might seem small, but it means a significant increase in accumulated heat, and that extra heat is driving regional and seasonal temperature extremes, reducing sea ice, intensifying heavy rainfall, and changing landscapes (Lindsey and Dahlman 2024). Not only are rising global surface temperatures influencing changes in weather patterns, they are also affecting human populations throughout the globe.

Some of the key points to remember from Chapter 1 will be important in understanding the issues discussed in this chapter. The main points were: (1) The temperature increases of 0.06 °C (0.11 °F) per decade since 1850 and the greater than three-fold increase in the rate of warming since 1982 of 0.2 °C (0.36 °F) per decade has been without a doubt caused by anthropogenic (human-caused) global warming (Eyring et al. 2021); (2) According to the National Oceanic and Atmospheric Administration, The National Aeronautics and Space Administration (NASA), and the World Meteorological Organization, the years from 2014 to 2023 ranked among the ten warmest years on record and 2023 was the warmest year since global records began in 1880 by a wide margin; (3) Researchers have statistically linked the water increase in storms (including

---

A coal-fired power plant in Shanghai, China. Coal is an abundant natural resource in China. Although the country has invested heavily in renewable energy development in recent years, it continues to be dependent on coal, burning more coal each year than the rest of the world combined. (Igor Groshev/Dreamstime.com)

hurricanes, mid-latitude cyclones, and snow storms), flooding, and droughts (that can lead to wildfires) to global warming. Regional extremes from seasonal temperatures on both the warmest and coldest record temperatures, as well as wettest and driest periods ever registered, also impacted communities, policy makers, and industries.

This chapter is divided into three main sections looking at the *problems*, *controversies*, and *solutions* that arise as nations commit to limiting global warming to below 1.5 °C or 2.7 °F and how they must face the already increase of 1.2 °C or 2 °F that has led to extreme weather events. The first part of the chapter, *problems*, looks at the specific problems that include the slow political and industrial progress toward substantial actions necessary to keep global warming from a continuous ever-increasing trajectory. The chapter also looks at the top nations emitting greenhouse gases and the top sectors that are powered by greenhouse gases. Specifically, it looks at the challenges top-emitting nations face when it comes to substantially reducing their carbon footprint as well as their progress toward their net-zero goals (where global greenhouse gas emissions from human activity are in balance with emissions reductions).

The second part of the chapter, *controversies,* discusses the vulnerable communities that suffer the harm of extreme weather events and the ethical considerations that the emitting countries must face as vulnerable populations face the effects of their emissions. Summaries of the world's most vulnerable populations are provided, which include developing nations as well as international demographic groups that are found in both wealthy and poor nations across the world. These vulnerable populations are identified by international relief groups as those unable to cope with the challenges of climate change and who are disproportionately affected by droughts, floods, storms, and wildfires due to a lack in resources and or a lack of ability to adapt to abrupt environmental change. This section also discusses the difficulties of assessing damage and loss from extreme weather events, the effectiveness of Nationally Determined Contributions in achieving climate goals, and the increase in destructive mining for minerals needed to implement green technology solutions.

Finally, the third part of this chapter, *solutions,* looks at proposed mitigation strategies for staying on track with the goals of the Paris Agreement (maintain temperatures from rising above 1.5 °C) by reducing the use of fossil fuels. The solutions discussed include green energy technologies, an improvement of natural carbon sinks, and last-case scenarios that would involve measures to alter the climate system through geoengineering.

## Problems

### Blurring the Truth by Creating Doubt

The international community of scientists including those at NASA, the World Meteorological Organization, the Intergovernmental Panel on Climate Change (IPCC), and tens of thousands of other atmospheric researchers agree—it is unequivocal that human actions since the Industrial Revolution have caused global warming. Through the burning of fossil fuels, humans have added to what is referred to as the greenhouse effect. Greenhouse gases are part of a natural component of the atmosphere that absorbs heat emitted by Earth's surface. The surplus of anthropogenic greenhouse gases has heated the planet at an unprecedented rate. The added heat has affected the water cycle, shifted weather patterns, and melted ice sheets (at a rate of 12 percent per decade) (NASA 2022).

The international scientific community agrees that these changes to the physical environment have contributed to extreme weather events. Extreme weather statistically attributed to climate change across the planet is becoming more intense and more frequent. The shifts in weather patterns that meteorologists have seen in the past few decades are now referred to as the "new norm," or in broader terms, "Climate Change." The news is full of stories reporting record-breaking heat waves on land and in the ocean, drenching rains, major snowstorms, severe floods, extended droughts, extreme wildfires, and hurricanes that stall and dump an exorbitant amount of rain like those discussed in Chapter 1.

Because extreme weather has made climate change tangible to many, one might think that current greenhouse gas levels have only recently been a concern. The truth is, scientists have studied the planetary influences of $CO_2$ and other heat trapping gases for centuries. Since the 1800s, scientists have understood that an influx of greenhouse gases could potentially increase global temperatures. Scientists such as Svante Arrhenius, Eunice Foote, Joseph Fourier, and John Tyndall, to name a few, began to publish, hold meetings, and warn governments of the possible consequences to unabated fossil fuel burning (Thompson 2019).

Politicians and fossil fuel companies have known for decades of the potential for greenhouse gases to cause significant global warming. In the 1960s and 1970s, environmental problems around the world brought issues like climate change to the forefront of the public's attention, spurring action. To address the issue of human-induced climate change, policy makers had to address the source of emissions. Studies pointed to the energy and transportation sectors that predominantly relied on fossil fuels as primary sources. This was no surprise to the large energy firms like Exxon, which had already studied the risk of increased emissions leading to a warming atmosphere. In fact, in 1981 an ExxonMobil scientist resigned after one of the company's spokespersons downplayed his finding that the company's emissions could cause unprecedented warming (Keane 2020). The average projected warming was 0.20 °C ± 0.04 °C per decade, which is consistent with the findings by independent academic and government projections published between 1970 and 2007 (Supran, Rahmstorf, and Oreskes 2023).

In fear of restrictions and regulations, ExxonMobil withdrew from climate change mitigation efforts and began the massive public-relations campaigns of the 1990s that supported inaccurate non-atmospheric science groups like the Global Climate Coalition. While the company said this was an effort to find non-partisan scientific evidence, the ads they promoted were not filled with scientific evidence or progress but were intended to create doubt. The ramifications of these efforts on public opinion and policy were far-reaching and linger today, as discussed further in the following section.

### *United States: Policy Pendulum*

The United States is no stranger to extreme weather events. Like the rest of the world, it is becoming more aware of climate change as cities battle with extensive, life-threatening floods, wildfires, and massive hurricanes, extreme cold spells, and record-breaking precipitation. Despite the evidence of extreme events, in the past few decades, the United States has swung like a pendulum when it comes to climate change policy.

Interestingly, the country was not always so bipolar or bipartisan on environmental topics. In fact, in the 1960s and 1970s, environmental catastrophes domestically and around the world sparked concern about the role of development of infrastructure without environmental protection. Rachel Carson's *Silent Spring* raised awareness about chemicals that had caused detrimental effects on nature and humans. Water pollution was so unregulated that the Cuyahoga River in Ohio caught fire, and the coast of Santa Barbara was contaminated with an insecticide spill. Americans of all political inclinations demanded protection of their lands.

Republican president Richard Nixon heeded the scientific data and the concern of citizens. In his 1970 State of the Union Address he declared an ambitious goal to restore natural resources by redirecting economic growth. Evident in Nixon's speech was that environmental protection was once a bipartisan priority. By the early 1980s, the science of human-induced climate change was already established. News outlets brought scientists' concerns into mainstream conversation. In 1988 NASA scientist James Hansen went before the Senate Energy Committee to advise the members. He informed them: "Global warming has reached a level such that we can ascribe with a high degree of confidence a cause-and-effect relationship between the greenhouse effect and observed warming." He added, "It is already happening now" (Krajick 2018).

At a campaign stop in Michigan in 1988, Republican presidential nominee George H.W. Bush (a self-proclaimed environmentalist), called for action on concerning environmental topics of the day, predominantly acid rain, deforestation, and climate change. Both Bush and his democratic opponent, Michael Dukakis, promised action during their campaign speeches. Once in office, Bush helped launch the United Nations Framework Convention on Climate Change, an organization that supervises global efforts to understand, mitigate, and adapt to global warming and climate change.

In the 2000 race, former Vice President Al Gore ran for president of the United States. He had been educating politicians and the public on global warming since the start of his career. He first became interested in the subject after taking a course at Harvard with Professor Roger Revelle, one of the first scientists to measure $CO_2$ in the atmosphere. During his presidential campaign, historical polls showed an increase in the number of Republican respondents who believed the media were overplaying the seriousness of human-induced climate change (Childress 2012). As discussed in the previous section, ExxonMobil had engaged in public-relations campaigns during the 1990s to sow doubt regarding the connection between global warming and human activity. Gore lost the coal-mining state of West Virginia (formerly a democratic stronghold). Voters reported voting for George W. Bush based on fears that Al Gore was a left-wing environmentalist who would eliminate coal jobs (as portrayed by Republican campaign adds). West Virginia was a loss that some say cost Gore the presidency (Bailey et al. 2022).

George W. Bush diminished action on climate change during his second term (ending in 2009). He rejected the Kyoto Protocol, the international agreement on global warming calling for all developed nations to diminish greenhouse gas emissions. Scientists reported that the administration pressured them to not speak about global warming. However, after Hurricane Katrina devastated the Gulf states in 2005, causing over $100 billion in damage and displacing more than one million people, the Supreme Court ruled that it was the federal government's duty to address greenhouse gas emissions. Responding to this, Bush issued more stringent fuel-economy standards and set targets to slow $CO_2$ emissions. "Many are concerned about the effect of climate change on our environment. Many are concerned about the effect of climate change policies on our economy. I share these concerns, and I believe they can be sensibly reconciled," he said in 2008 ("President Bush" 2008).

Despite George W. Bush's efforts to curtail emissions, misinformation about climate change continued into the next decade. Almost a decade later, politicians continued to argue about the truth connecting fossil fuels to global warming. Government officials tried to clarify the matter. In 2018, the attorney general's office of the state of New York launched a three-year investigation and filed a lawsuit against ExxonMobil. It represented the most significant legal effort to hold a fossil fuel company and its chief executives responsible for misleading the public about the threat of future climate change from its industry. Their findings concluded that oil company scientists had reached a consensus in the early 1990s that human-caused climate change was measurable and that the planet was already seeing the projected increases in temperatures (Schwartz 2018).

The idea that climate science was up for debate became part of mainstream culture. Conversations used the word "believe" when referring to the science instead of "have you read/understand" the science. The public aligned climate science in the same categories as topics that you did not have to prove or study. Scientists were confused by the public's choice to ignore data or science in general and to turn toward affiliations. Even meteorologists often avoided the topic, as the truth of the matter would cause angst among the public (CNN 2019).

Partisanship continued to divide the issue of climate change while President Obama occupied the White House, despite storms like Hurricane Sandy, one of the most intense and damaging East Coast storms of all time, whose winds extended 1,000 miles (three times that of a typical hurricane). Perhaps this extreme weather event helped the public find a need to understand the threat of a possible increase in extreme weather events. Instead of aiming to secure bipartisan support for climate change policy (which was likely to not pass), Obama instructed his agencies to create new regulations. During his presidency, the Environmental Protection Agency (EPA) issued nearly 4,000 rules, according to a House Energy and Commerce Committee report. Obama's hasty measures angered Republicans who saw the extensive legislation as an attack on businesses.

Shortly after the extensively damaging southwestern droughts (2012–15), Donald Trump campaigned with the idea that environmental protection, including the limit of greenhouse gas emissions, was an attack on the freedom of American people. After his election in 2016, Trump surrounded himself with like-minded individuals who would be in charge of the United States' land, water, and air, policies. For example, he appointed Oklahoma attorney general Scott Pruitt as head of the EPA. The public raised concerns about Pruitt's skewed view of science and denial of the evidence of anthropogenic climate change.

During Trump's first presidency (2017–21), the world witnessed extreme hurricane weather including Irma, Maria, and Harvey in 2017, and Hurricane Michael (2018). The public, reeling from these storms, pointed out Pruitt's misunderstandings of the effects of global warming, citing an article for the *National Review* that he had coauthored, incorrectly stating that the "debate is far from settled. Scientists continue to disagree about the degree and extent of global warming and its connection to the actions of mankind" (Malakoff 2016).

Despite Trump withdrawing the United States from the Paris Agreement in 2017, thousands of leaders nationwide, including governors, mayors, universities, private corporations, and individual citizens, stepped in to fill the void created by the lack of federal climate leadership. American participation in initiatives such as America's Pledge, the United States Climate Alliance, We Are Still In, and the American Cities Climate Challenge displayed the people's commitment. The organizations aimed to stay on task and accelerate efforts to tackle climate change at the local, regional, and national levels. In 2017, The House Climate Solutions Caucus had fifty members, twenty-five from each party, and some of the more controversial measures slowing action on climate change were halted as a result of Republicans who understood the scientific evidence and remained concerned.

Science articles of 2021 confirmed that the 0.25 °C warming since 2010 had led to the emergence of heat extremes that would be virtually impossible without anthropogenic global warming. Record rainfall extremes worldwide increased with one in four rainfall records in the last decade attributed to climate change (Robinson et al. 2021). With no time to waste, the United States rejoined the 198 nations that signed the Paris Agreement when President Joe Biden took office in 2022. On his first day of presidency, he set goals for the United States to reach net-zero emissions, economy-wide, by no later than 2050. He hoped to achieve that goal through multiple cost-effective pathways, including deploying carbon pollution-free electricity generating resources, carbon capture initiatives, and the use of existing nuclear energy. The federal government under his administration passed the Inflation Reduction Act in 2022 with incentives to build US renewable energy infrastructure and cut greenhouse gas emissions by roughly 40 percent below 2005 levels by 2030. The act contained hundreds of billions of dollars in tax credits and spending for clean energy technologies like wind turbines, solar panels, batteries, electric vehicles, and hydrogen fuels.

The current popularity of voting along party lines without much reflection or questioning of facts versus propaganda has made it increasingly difficult for governments to enact climate policy. The US federal government flips like a pancake on climate initiatives. However, experts at the Resource for Future point out that while a disparity along party lines exists, when it comes to realization of the effects of climate change in the future, voters are still interested in environmental policy that mitigates or relieves the effects of extreme weather such as drought, wildfires, hurricanes, or other hazards regardless of their understanding of its relation to global warming. People want federal action on disaster adaptation especially at the state level where local leaders know the needs of their constituents.

## Philosophical and Cultural Challenges

Researchers at Yale University have found that people's understanding of climate science is shaped in large part by cultural attitudes driven by political philosophies or religious beliefs that also impact how they feel society should respond to climate change. Opposing philosophies such as egalitarianism and individualism play a part in hindering climate action. Egalitarians follow the doctrine that all people are equal and deserve equal rights, opportunities, and consideration. For many people, the philosophy also includes the natural world (plants, animals, and the environment). The egalitarian school of thought is part of the political philosophy of social equality for all people. Egalitarians are concerned with extreme weather events around the world, as those without access to healthcare, insurance, and social support systems are at an unfair disadvantage. They also tend to favor regulations that provide support across various socioeconomic groups (Carman et al. 2020). When it comes to the impact of extreme weather on various populations, egalitarianism becomes linked to the idea of climate ethics. In contrast, people who support an individualistic philosophy support individual liberties, autonomy, and opportunity with an emphasis that people should worry about their own problems, do as they like, and not be extensively governed by political institutions to ensure the needs of others.

Opposing views becomes a roadblock to climate action because any action on climate change includes individuals, nations, and corporations. That is perhaps why it is referred to as one of the most challenging issues facing the world today. Some have even described it as a "perfect moral storm" based on three things. The first is that it is a global issue with catastrophic consequences no matter where emissions come from, and second the conundrum that while most nations as a whole agree that emissions must be regulated and lessened as soon as possible, individuals within nations do not want to limit their own personal emissions (Gardiner and Hartzell-Nichols 2012). Last, those who are most vulnerable to extreme weather events are those who have emitted the least.

Social scientists say that in order to reach a *social* consensus, people's opinions, ideologies, personal experiences, psychology, and the influences of those they identify with, must be considered. Not only that, but people's inherent hierarchy of needs must be met. According to prominent American psychologist Abraham Maslow, who created a pyramid to illustrate an individual's hierarchy of needs, a person is not interested in satisfying the needs of others until their basic needs are met (English 2021). This was evident in research from the early-2000s when Maslow's theory was supported when the pursuit of wealth outcompetes climate action. Google searches between 2004 and 2010 were centered on employment and financial assistance, not on climate change topics. The searches were analyzed by economists who concluded that recession increased concerns about unemployment at the expense of climate change and may be correlated with people's denial of global warming's existence (Kahn and Kotchen 2011). This conclusion has been contested due to the fact that Google searches cannot be directly representative of the US public as a whole or the extent of their concerns toward climate change.

## Psychological Impacts of Extreme Weather

Is the proof in the pudding? Does experiencing extreme weather events override tribalism or political affiliation to scientific topics? Recent polls have shown that the US public is growing in acceptance that extreme weather events like the massive atmospheric rivers hitting California bringing massive floods, snow accumulation amounts in higher elevations, and coastal landslides have something to do with climate change. The heatwaves of 2023 and the massive fires in early 2024 in Texas just may have influenced opinion.

Thanks to attribution science, the public can see for themselves the added likelihood that the events at hand are enhanced by warmer global temperatures. A recent Gallup poll found that one-third of all Americans have been exposed to extreme weather events since 2020. According to *New York Times* reporter Derrick Bryson Taylor, based on interviews conducted with adults living in all fifty states and Washington, DC hurricanes and winter weather, such as snow, ice storms and blizzards, were the most common extreme weather events cited, followed by extreme heat and floods (Taylor 2022).

People in the path of extreme weather events also showed a shift in their concern for continued global warming. The Gallup poll indicated that those who experience extreme weather tend to see climate change differently than those who have not. Researchers found that those with personal experience of an extreme weather event had a higher concern (63 percent) with future global warming compared to those who had lower concerns and had not experienced extreme weather events personally (33 percent). Of those who lived through an extreme weather event (67 percent) said that the government was not doing enough about climate change as opposed to 48 percent who had not. The personal experience of a weather-caused disaster may promote political change in climate change mitigation and adaptation policy.

Steve Cohen, who writes for the Columbia Climate School, proposes that extreme weather events should be planned for and not require action or funding after the event. Policy and relief operations must be in place so that homelessness and loss of income, education, and health are not substantially disrupted while governments argue about policy. He brings up the point that to be resilient after extreme weather events, we must consider the interconnectedness of our societies, globally to locally. Modern industry and homes are interconnected by energy needs, roads, emergency personnel, water conduits, and sewers (Rudiak-Gould 2013).

Researchers have been studying and documenting their findings about the public's growing concern about global warming, extreme weather, and climate change and the lack of political, individual, and/or corporate mitigation efforts. Mental health clinicians are also reporting seeing an increase in patients with symptoms of climate change anxiety—also referred to as eco-anxiety, eco-grief, or climate doom. The trend also follows conclusions from public surveys and data from internet searches. For example, Google searches for "climate anxiety" soared by 565 percent in 2021, according to the news outlet *Grist* (Moench 2023).

## International Efforts at Lowering Emissions Are Slow at Best

Most governments, fuel companies, and even countries whose main industry are fossil fuels (like Saudi Arabia) have finally agreed to reduce greenhouse gas emissions. Despite their current agreement that action needs to be taken, it may be a bit late. Studies reviewed by the IPCC reveal that the actions thus far taken by nations who pledged to uphold the Paris Agreement (an international treaty to keep warming below 1.5 °C) are too minimal to curtail global warming and avoid the most dangerous of climate impacts. According to Our World in Data, growth in emissions was still relatively slow until approximately 1950. From 1950 to 1990 global emissions of $CO_2$ quadrupled from six billion tonnes to twenty billion tonnes. Emissions continued to grow rapidly, currently global emissions average nearly thirty-five billion tonnes each year. The level as of 2024 accounted for an increase of over 149 percent since pre-industrial levels. Emissions growth has slowed over the last few years, but they have yet to reach their peak (Ritchie and Roser 2020).

According to the World Resources Institute, the top three emitting countries, China, United States, and India, contribute fifteen times the greenhouse gas emissions of the bottom 100 emitters. China and the United States are responsible for 30 percent (10,065 million tons of $CO_2$) and 14 percent (5,416 million tons of $CO_2$) of emissions respectively (Friedrich et al. 2023). The top ten emitters—China, the United States, India, the European Union, Russia, Japan, Brazil, Indonesia, Iran, and Canada—emit primarily through the energy sector whose emissions have increased by 62 percent since 1990. This includes emissions from the generation of electricity and heat as well as from buildings, transportation, manufacturing, and construction. Other major contributing sectors include agriculture, industrial emissions, land-use and forestry, and waste, in that order. Land-use change and forestry (decrease of 14 percent), is the only sector with a decrease in overall emissions. All other sectors have continued to increase emissions since 1990, including agriculture (16 percent increase), industrial emissions (203 percent increase), and waste (19.5 percent increase).

### *China: Coal Dependence and Renewables Investment*

In the early years of climate change initiatives (1990s to 2000s), China argued that as a developing economy, it should have less of a requirement to curb emissions due to the fact that developed nations had been emitting at high rates without restrictions for decades. However, the tangible threat of recent extreme weather events has encouraged the government to join the international community in the quest for a swift transition to clean energy. In 2020, Xi Jinping, the leader of China, in hopes for his country to be seen as a modern superpower, declared to the world that it would transition to clean energy at record-speeds and with the latest technology. During a speech to the United Nations General Assembly that year, and after devasting floods on the Yangtze River, Mr. Xi said, "Humankind can no longer afford to ignore the repeated warnings of nature." It was then that Xi announced China's plan to become carbon-neutral before 2060 (Bradsher 2023).

In recent years, China, like the rest of the world, grappled with extreme weather events that were the cause of lost lives and extremely costly. According to a story in *CNN World*, in 2022, record-breaking rainfall brought severe flooding and landslides to large regions of southern China. The floods killed dozens of people, displaced millions and caused billions of yuan in economic losses. That summer, the extreme amount of rainfall broke "historical records" in coastal Fujian Province, and parts of Guangdong and Guangxi provinces. At the same time, a heat wave began to envelop northern China, with temperatures over 40 °C (104 °F). Over 900 million people—or about 64 percent of the population—was affected. The highest-level red alerts were issued for eighty-four cities. For example, three cities in the central province of Hebei and one in Yunnan in the southwest recorded temperatures reaching 44 °C (111 °F), according to the National Climate Center (Gan 2022).

Although Mr. Xi invested in efforts for renewable energy development, the global pandemic and extreme weather events including the heatwaves lasting over seventy days in 2022, required the generation of energy quickly and affordably for over one billion people. To fulfill the unexpected need, China increased coal consumption in the face of economic hardships. Unfortunately, $CO_2$ emissions from energy production, cement making, and other industrial uses have increased since 2019 and more construction permits for coal-fired power plants were granted in the first six months of 2020 than in 2018 or in 2019. Despite this, Mr. Xi in April of 2021, said that the government would strictly control the growth of coal consumption through 2025 and then "gradually reduce it" through 2030.

With its abundant underground coal reserves larger than any nation, energy officials said coal was essential to China's energy security. Zhang Jianhua, director of the government's National Energy Administration in 2023, described coal as the "ballast stone" of his country's energy mix. The continued use of coal and development of new plants may affect the pace at which China transitions to renewables. Other nations fear that China will not be able to meet the lofty goals declared at the 2020 UN General Assembly. These included the addition of 1,200 gigawatts of renewable capacity, to have $CO_2$ emissions peak before 2030, and to reach net-zero by 2060 (Zhang 2023).

The excavation of coal in China is also a double-fold problem for the country's tally of emissions. As coal is extracted from underground reservoirs, methane (a greenhouse gas twenty-eight times more potent in warming potential than $CO_2$) is released to the atmosphere. China's physicists estimated that a quarter of all methane emissions in China come from its more than 100,000 coal mines, many of which have long been abandoned but still emit methane.

From 2021 to 2023 China held a national campaign to expand coal mines and build more coal-fired power plants. In the spring of 2023 thousands of workers toiled without leave to finish one of the world's largest coal-fired power plants in southeastern China (Bradsher 2023). Around that time, China's operating capacity of coal power plants was 1,000 GW, compared with 227 GW in the United States, 232 GW in India and 118 GW in the twenty-seven countries of the EU and the United Kingdom (IEA 2022). Also in 2023, like most countries, China grappled with the desperate need to derive enough energy during the hottest year on record. Energy was needed to power millions of air conditioning units. The only reliable source at the time was coal. When it comes to coal in particular, China burns more every year than the rest of the world combined. As a result, China continued to develop the industry for energy security.

Much like the United States, China's highest amount of solar and wind potential lies in the western deserts of the country where sunlight and wind are significantly more available. These areas are extremely far from the highly populated cities of the eastern portion of the nation, requiring extensive infrastructure to connect the power generated at solar and wind farms to the populated zones. More miles of these lines have been built in China than anywhere in the world. The lines are astronomically expensive. Each line requires hundreds of miles of 200-meter-wide strips of land that must be purchased from a private owner if it is not already government land.

On a positive note, China has become the world's leader in renewable energy. Experts estimate that China is set to meet that target by 2025 and could boast as much as 1,000 gigawatts of solar power alone by the end of 2026, an achievement that would make a substantial contribution in meeting the 2030 targets of the Paris Agreement. Despite the increase in coal plants, fossil fuels now make up less than half of China's total installed generation capacity, a dramatic reduction from a decade ago when fossil fuels accounted for two-thirds of its power capacity (Hilton 2024). In 2023, Chinese climate envoy Xie Zhenhua and US climate envoy John Kerry shook hands on a pledge to triple renewable energy globally by 2030 (Hilton 2024).

According to data from Scimago, a portal that analyzes scientific indicators of journals and countries using the Scopus database, China leads the world in terms of scientific research output in renewable technologies. In 2020, China produced almost three times as many papers related to renewable energy as the United States (Otto 2019). As early as 2018, China produced 57 percent of the world's patents in renewable energy according to data from Irena.org, an intergovernmental organization that supports countries in their transitions to a sustainable energy future. A report by the Climate Bonds Initiative noted that by 2021, China was the fourth largest issuer of green bonds, which allow companies to raise funds from investors for renewable projects including wind and solar. These also came with additional government tax incentives for investors (Climate Bonds Initiative and China Central Depository & Clearing Research Centre 2021).

### *India*

In 2020, India's Ministry of Earth Sciences published the first climate change assessment report for the country based on data from 1901 to 2018 and showed that the country's average temperature in that period had risen by roughly 0.7 °C. In a 2023 article in the journal *Nature*, scientists report that India is facing a harsh new reality. In southern cities like Kerala, floods frequently destroy farmlands and homes. In the sea off Mumbai, fewer fish are caught owing to an increase in cyclones and heavier rainfall. Northern cities like Rajasthan face record-breaking temperatures and cattle farmers struggle with the unbearable heat and resulting water shortages. In Uttarakhand, forest fire frequency has increased. And coastal villages off the Bay of Bengal are experiencing rising sea levels and coastal erosion that have forced hundreds of families to leave (Hall 2023).

The *Nature* article includes a chart showing the increase in extreme weather events with a summary that concludes that since the pre-industrial period, the 0.7 °C increase in temperatures is evident in retreat of the Himalayan glaciers, worsening droughts, increase in flash floods and landslides, and the intensification of cyclones across the country's 7,500-kilometer-long coastline.

These events are now common. In the first nine months of 2022, for example, India had an extreme weather event nearly every single day. "It can be a year-long swaying of the climate sledgehammer. We now have disasters in pretty much every season," reports Raghu Murtugudde, a climate researcher at the Indian Institute of Technology Bombay (Hall 2023).

With personal interest and global intentions, India reminded the world in their address to the United Nations Framework Convention on Climate Change that they represent a culture that calls their planet Mother Earth and whose ancient texts declare, "Keep pure! For the Earth is our mother! And we are her children!" India updated their Nationally Determined Contributions to be net-zero by 2070. The subcontinent continues to pledge to generate 50 percent of its electricity from renewable energy sources by 2030 and to decrease their cumulative emissions by one billion tons, while decreasing the emissions intensity of its GDP by 45 percent.

India's commitment to achieving a net-zero target by 2070 would require various strategies to diminish its reliance on a fossil fuel-based energy grid. The transition to renewable energy would need to incorporate energy sources such as solar, hydrogen, and biomass derived fuels. They would also need to decarbonize the generation of thermal power, cement, metal, and their petrochemicals. They do request that while they are committed to developing sustainably, they need the developed and resource rich countries to reduce the burden of their action from being borne by developing countries.

A report titled "Getting India to Net Zero" estimates that India will require approximately USD 10.1 trillion to meet its 2070 target. "It is certainly a challenge for the country to meet its economic goals, and to expand reliable and affordable energy access to its citizens while staying on course in achieving ambitious climate targets," said Harjeet Singh, Head of Global Political Strategy at Climate Action Network International. Singh also mentioned India's concern for the poor and his opinion that they must be incorporated into plans for future sustainable development. US special climate envoy John Kerry hopes that financial aid to countries like India from the United States will set an example for promoting rapid clean energy transitions and economic development in emerging economies around the world (Asia Society 2022).

According to the International Energy Agency (IEA), India's energy demand will be the highest globally during the 2020 decade, rising by over 3 percent per year. India is also projected to have the world's largest population on Earth by 2025. They will have exorbitant energy demands and an increased demand for the expansion of urbanization and industrialization. Currently, over half of India's energy supply comes from coal. To provide power for their growing population, India is expected to increase their coal-based power supply by 35 GW by 2030. The IEA acknowledges that India's primary challenge is to satisfy the growing demand for electricity with renewables and reduce the reliance on coal. Balancing energy security with green growth was a key topic at the recent Energy Transitions Working Group meeting under G20 in Bengaluru. India is hoping to be a USD thirty trillion economy by 2050. This economic growth will add an immense energy demand for each and every one of its citizens. Economists predict that India's transition from fossil fuels to renewables will be gradual. Analysts at the meeting of the G20 concluded that India's reliance on fossil fuels are likely unavoidable for a few more decades until the country secures its economic stability.

India is actively exploring carbon capture, utilization, and storage (CCUS) initiatives, especially in industries that are difficult to overhaul in a short period of time. These industries include thermal power plants and those involved in the production of steel, cement, aluminum, ammonia, and petrochemical industries. Private corporations are also keen on developing large-scale CCUS plants. India hopes to model some of the methods already successful in Norway, China, and the United States. Policy makers view carbon capture initiatives as a top mitigation strategy that could fit into the existing coal infrastructure and help avoid major economic or social disruption as the country has relied on coal for affordable energy, income, and social support for the poor. Although CCUS has substantial mitigation potential in deep decarbonization scenarios in India, efforts may be halted without financial support (Shaw and Mukherjee 2022).

Despite India's challenge to support the needs of the general growing population, the demand for urbanization, and the challenges of its poor citizens, the country has made large strides toward their clean energy goals particularly in the solar energy sector. Solar photovoltaic technology is reducing the country's reliance on coal. The 300 days of clear sunny weather annually have the possibility of generating approximately 5 quadrillion kilowatt-hours (kWh) of solar energy per year. Experts say that this amount of energy exceeds the country's total fossil fuel energy reserves if harnessed for a year. The electricity directly generated by photovoltaic cells can be connected to the conventional grid to offset the use of coal-fired power. India developed large on-grid solar projects called solar parks, with the biggest ones located in Bhadla (the world's largest solar park), Pavagada (the world's second largest solar park), Kurnool (world's seventh largest solar park), and Rewa (the world's twelfth largest solar park). India has approved sixty-one new solar parks that would add 40 GW. The largest challenge for India in expanding their solar power is the challenge of finding land. However, they have been successful in adding floating solar plants. The world's largest floating solar plant of 600 MW is underway in Khandwa, Madhya Pradesh.

Hydrogen energy can be labeled a variety of colors (green, blue, brown, pink etc.), depending on the source, production methods, and resulting byproducts. Green hydrogen like the name implies is the only type produced in a climate-neutral manner making it critical in the race to reach net-zero by 2050. Green hydrogen is essentially hydrogen produced by splitting water into hydrogen and oxygen using renewable electricity. The hope is that green hydrogen will help in the transition to clean energy technologies and to decarbonize not just the electricity sector but also other hard-to-get-rid-of sectors like fertilizer production and petroleum refining. Experts think that India's government policies should further boost supply of green hydrogen and support the sustainable development and implementation of a green hydrogen consumption mandate for fertilizer production and petroleum refining. India has an aggressive target of five million tons a year of green hydrogen capacity by 2030. This may come to fruition with the financial commitments of major developers. At the World Economic Forum Annual Meeting in Davos, Indian renewable energy developer ACME signed a memorandum of understanding worth $7 billion with the state government of Karnataka to develop an integrated solar to green hydrogen to green ammonia. The goal of the new facility will be to produce 1.2 million tons a year of green hydrogen by 2027. In June, French oil and gas giant Total Energies announced a partnership with Adani New Industries of India to invest $50 billion over ten years to produce green hydrogen.

The Institute for Energy Economics and Financial Analysis reported a record-level increase in the past few years for India's renewable energy grid including a $14.5 billion investment in renewables in the 2021 to 2022 financial year (an increase of 125 percent). The government understands that energy storage will be a key component to achieving India's renewables targets. Initiatives and policy support for all their sectors is at play and the combination of solar, CCUS, wind, and the development of green hydrogen may help India reach their National Determined Contributions.

# Controversies

## The Poorest Nations with the Least Emissions Are Most Vulnerable

Observations by social scientists and relief organizations show that the ability to prepare and protect oneself from health concerns, loss of property, and mental illness pre and post extreme weather events is hardest for those in developing nations, and in vulnerable populations. Households with single parents, tribal communities, children, the disabled, ethnic minorities, the homeless, migrant workers, the elderly, and socially marginalized groups are most at risk regardless of their country's wealth. Many of the people who are part of these groups are vulnerable because of a combination of factors that will be discussed in this chapter including geographical location, financial stress, cultural/political challenges, and access to resources (EPA 2021).

According to Mia Amor Mottley, the prime minister of Barbados, a country struggling with the impacts of hurricanes, 40 percent of the world's population live on the frontlines when it comes to extreme weather. She argues that while heatwaves and fires affect North America and Europe, the impacts are erased in a matter of months due to their financial status and climate resilience (the ability to return to a normal life after an extreme weather event or an environmental catastrophe linked to anthropogenic climate change of wealthier nations). Mottley argues that extreme weather events are impacting 189 million people per year in developing countries (Alleyne 2019).

As global warming fuels extreme weather events that impact societies disproportionately, social justice issues arise. Clearly the suffering of people in the poorest countries worsens as richer countries continue to emit greenhouse gases scientifically linked to the increase of extreme weather events that further create inequality. Many poor nations as well as supporting activist groups believe that heavy-emitting countries and companies have an ethical duty to compensate those affected by extreme weather and climate-related events resulting from emissions. For example, a study by the Loss and Damage Collaboration concluded that six fossil fuel companies could have paid for the cost of extreme weather event related damage in developing countries and still have made a $70 billion dollar profit. In the report, researchers also show that fifty-five of the most climate-vulnerable nations have had economic losses totaling over half a trillion dollars from 2000 to 2020. The changing climate affects all people to some degree, but some communities are better able to prepare, cope, and adapt to global warming and the resulting extreme weather events (Walsh and Skeaping 2022).

The ND-GAIN Index ranks 181 countries, calculating a country's vulnerability to climate change and other global challenges as well as their climate resilience. The more vulnerable a country is the lower their rank is on a list from 1 to 181. For example, Norway has a score of 1 and is the least vulnerable country to climate change. Chad, Somalia, and Afghanistan are the top three most vulnerable countries; incidentally, they are also listed among those that contribute the lease greenhouse gases. A common theme is seen among the most vulnerable including a lack of infrastructure, geographic exposure, and internal political conflict.

### *Chad*

Chad, with an ND-GAIN score of 181, has been identified as the most vulnerable country to the effects of climate change. Eighty-seven percent of the country's citizens, most of which live in rural zones, are classified as poor according to the Multidimensional Poverty Index, an index that incorporates a person's health, education, and living situations. Chad also has frequent conflicts, including an ongoing civil war for thirty-five out of the fifty-seven years since it gained independence from France. The country struggles to share its limited resources with nearly 380,000 refugees from Darfur (a region in western Sudan), and Central African Republic (southern bordering nation). Also adding to the troubles of this nation is Boko Haram, an Islamist militant organization based in northeastern Nigeria, which creates conflict and civil unrest.

To help alleviate Chad's poverty, the country began producing petroleum in 2003, which now accounts for 93 percent of its exports. Chad will need massive financial aid to transition from petroleum and other fossil fuels. Because the majority of its population is still involved in agriculture, developing sustainable agriculture and livestock farming practices in providing food security and employment for its young population is essential.

Chadians are struggling with temperatures that have been increasing since the 1950s. Drought has also been increasing, and 90 percent of the country's largest lake, Lake Chad, has disappeared over the past fifty years due to a combination of droughts and increasing withdrawals for irrigation. Climate change projections indicate that temperatures will continue to rise, and aridity will continue and spread throughout the twenty-first century, resulting in lower crop yields, loss of pasture land, and a harder life for farmers and pastoralists who dependent on Lake Chad. The urban areas are challenged by the influx of people from the zones that are no longer arable. According to the World Bank, cities struggle with poor sanitation services like sewage, storm water drainage, and waste collection. If the cities were to flood as they did in 2010, 2011, and 2012, the lack of sanitation infrastructure could infect the water supply and lead to cholera outbreaks. The impact of climate-related disasters such as droughts or floods becomes magnified when a country does not have the resources to combat them.

The young and often unemployed population of Chad has been at the front of flooding disasters. Massive rainfall amounts led to unprecedented flooding in August of 2022, affecting more than 340,000 people living in 55,000 households (100,000 more people than the floods of the previous year) according to a report by the UN humanitarian coordination office, OCHA. Researchers from the International Organization for Migration, are using Displacement Tracking Matrix, to develop data collection mechanisms to capture the extent of the inundation and number of victims, including those displaced. They hope the data will clarify information on the impacts of floods on populations and key infrastructures such as houses, schools and health centers to better inform the humanitarian response.

Extreme weather patterns including increased storms and a higher degree of variability (which causes stronger flash floods) highlight the adverse effect of climate change in the region. Not only is global warming creating a large risk from floods but a combination of long-lasting heatwaves and a weak monsoon season threatens the agriculture sector. Heatwaves and drought have been common in the past few years, and researchers think this may lead to the reduction of photosynthesis and leave crops vulnerable to pests and disease. The high evaporation rate during times of heatwaves contributes to soil degradation, soil compaction, and a lack of moisture threatening the health of microbes in the soil. All these factors hamper agriculture productivity. Heatwaves also hurt the pastoralists and their practice of livestock rearing, overall threatening food security. Animals also fall prey to heat stress impacting their growth, fertility, and productivity.

Chad clearly requires substantial aid from developed nations to adapt to a changing climate. This would include climate-smart agricultural practices with soil and water conservation methods and pest management. Chad would also need to establish water storage facilities, reservoirs, irrigation systems, and rainwater harvesting devices to help the agricultural sector survive during extended periods of drought. Better meteorological stations and warning systems could deliver timely weather information to help citizens (especially those that rely on the land) to make better decisions for the timeliness and protection of agriculture and livestock.

### *Somalia*

Somalia, with an ND-GAIN score of 180, ranks as the second most vulnerable country to climate change. Located in the Horn of Africa, Somalia's climate is predominantly hot and arid apart from two rainy seasons that can deliver up to 700 mm in the south to less than 100 mm in the very northeast. Mean temperatures in Somalia are among the highest on the globe. Somalia is at a high risk to famine. Sixty percent of Somalia's 15.8 million people are pastoralists or semi-nomadic. Weather dictates crop yields, and those in turn dictate the economy (agriculture accounts for nearly 65 percent of the GDP). Droughts and floods can be devastating.

The United States Agency for International Development's (USAID) report on Somalia found that global warming has increased Somalia's temperature by 1.5 °C since 1991. Extended droughts have become more common in the past twenty-five years, and severe droughts like the two that devastated the country in 2011 and 2017 are now a common risk. Projection models show that drought and floods are likely to increase in intensity and frequency. Models have also projected that monthly rainfall will increase slightly from September to December between 2040 to 2060 which may result in flash floods due to dry compacted soils. The average temperature is expected to increase by 1–1.75 °C from 2040 to 2060, and projected to increase by 3.2–4.3 °C by 2100. With an extended arid landscape and water insecurity, more and more Somalians are displaced annually (USAID 2023).

However, it is not just the country's aridity that poses a threat when extreme weather events arise. The UN envoy for Somalia reports that the extremist group Al-Shabaab and conflict in the Laascaanood region continue to cause civilian casualties and humanitarian crisis. Prolonged aridity and the resulting loss of water and water dependent resources increase conflict stemming from the fight over the resource. Somalia needs to implement emergency management systems that respond to immediate needs by individuals and communities and ways for people to rebuild after storms or drought devastate an area or conflicts arise over resources.

### *Afghanistan*

Afghanistan, with an ND-GAIN score of 179, is located at the meeting of Central, West, and South Asian regions. It is in the interior of the Asian continent and is mountainous with peaks reaching to 7,000 m in the Hindu Kush region. In the southwestern region, low elevations regions extend across the southwestern plateau. The Berkeley Earth data set reveals an average increase of 1.5 °C between 2000 to 2017. Afghanistan's Second National Communication to the UNFCCC in 2019 reported increased frequency of hot days and nights especially in the central and southwestern regions. The northeast has also experienced glacier loss.

Afghanistan is among the world's poorest nations with a national poverty rate greater than 50 percent. Most of the undernourished population lives in an arid rural zone subject to ongoing conflict and political instability. Afghanistan has high unemployment rate (42.5 percent) and a heavy reliance on agriculture (25.8 percent of its GDP). Nationally Determined Contributions submitted in 2016 highlighted the nation's very modest contribution to the causes of climate change and their need for financial resources to adapt to the diverse hazards brought about by extreme weather events.

Climate models for Afghanistan show a trend of consistent warming across different emissions greenhouse gas concentration scenarios. Projected precipitation trends indicate a slight reduction in average daily rainfall, but an increase in intensity for extreme rainfall events which could threaten the country with flash flooding events. Most climate models show that Afghanistan can expect levels of warming above the global average with a worst-case average rise of 5.5 °C by the 2080s and the 2090s, under the highest emission pathway. This is significantly higher than the projected global average rise of 3.7 °C above the 1986 to 2005 baseline for the same highest emission pathway. A large difference is seen in the rise projected under the lowest emission pathway of 1.4 °C. Extensive reductions in global greenhouse gas emissions by the United States, China, and India and other top emitters would make a profound difference. Such efforts might also protect Afghanistan from even more extreme projected rises in minimum and maximum temperature.

## Class and Vulnerability

Within discussions of vulnerable populations, most people think of those in developing countries who are often portrayed on world news in far off lands with lives whose infrastructure most of us cannot relate to. However, the same demographics within developed nations are more susceptible to suffering from extreme weather than the average population. These include Indigenous peoples, ethnic and racial minorities, migrants, children, and people with disabilities and single parents. Indigenous populations are exceptionally vulnerable due to the fact that they are often not part of the international climate change governance because they are often part of independent native alliances. The material constraints and the designation of Indigenous peoples as nonstate observers also poses challenges for participation in government assisted programs. Lower-income population groups experience the greatest adaptation gaps linked to socioeconomic factors such as access to financial aid, knowledge about adaptation policies, population density, unemployment rates, as well as physical risks such as those that discontinue needed resources such as health care, childcare, and transportation. Political factors like ineffective leadership or corruption also add hurdles in adaptation efforts (Dang, Hallegatte, and Trinh 2023).

## Assessing Loss and Damage

"Loss and damage" is a phrase used to describe the impacts that climate change and extreme weather events have on populations. Loss and damage assessments come into play, for example, in climate litigation following extreme weather events. The IPCC has categorized loss and damage into two groups. The first is any loss that involves income and physical assets. The second group includes losses that are non-economic including mortality, mobility, and the loss of wellbeing. Non-economic losses can be difficult to quantify in monetary standards and can often be irreplaceable.

A scientific study published in 2019 set out to systematically analyze climate-related, non-economic loss, also referred to as "intangible loss." These losses included what people value that cannot be replaced, keeping in mind the variation in the importance of objects, places, and traditions. The study aimed to understand where damage was happening due to climate change, what people valued most and why, and if the damage to reported loss was accepted or not by those who suffered the loss. Intangible loss was reported in over 100 documents and included systemic loss to biological systems including the loss of biodiversity and species, physical health, and human life. Psychological harm was also reported by groups whose loss from climate change events threatened their cultural practices, traditions, heritage sites, homes, civil rights, travel patterns, loss of sense of order in the world, sense of place, social fabric, and ability to govern oneself. Although the loss of intangible things is felt across the rich and poor, most reports of loss come from wealthier nations, while poor nations had a smaller or no record of loss (Tscharkert et al. 2019, 58–72).

## Nationally Determined Contributions

In 2018, the IPCC's Special Report: Global Warming at 1.5 °C concluded that limiting warming to 1.5 °C instead of 2 °C would prevent a greater amount of poverty, extreme heat, sea level rise, habitat loss, and drought. However, the task of reducing greenhouse gas emissions significantly enough to keep temperature rise below 1.5 °C requires a substantial economic and social evolution, and the best way to get there is a controversial topic.

Within the 32-page document of the Paris Agreement is the framework for global climate action, including the mitigation of and adaptation to climate change. When a country signs the Paris Agreement, they concede to monitoring, verifying, and reporting their progress toward emissions-reduction targets and be willing to have outside experts evaluate their success. The plans for action are known as Nationally Determined Contributions (NDCs) or Intended Nationally Determined Contributions. Countries are also expected to review their pledges and commit to establishing stronger targets every five years. They must also participate in a "global audit" that measures collective efforts toward meeting the Paris Agreement's long-term goals. Developed countries also have to estimate how much financial assistance they will allocate to developing nations to help them reach the goals and requirements of the Paris Agreement.

However, there are no financial penalties for failing to meet their pledges, and commitment is based on an honor system and an ethical responsibility for the greater good. Thus, the effectiveness of NDCs in achieving climate goals is debated, as progress is considered by many to be too slow. NDCs for 2024 are modeled to increase warming by 3.2 °C (5.7 °F) by 2100, overshooting our 1.5 °C/2 °C target, but new developments like net-zero pledges may lower that projection. Avoiding the worst extreme weather events will require reversing the upward trend in all sectors and rapidly decreasing emissions to net-zero by 2050.

The Paris Agreement requires one of the largest global transformations of infrastructure and societal behavior ever undertaken. Fortunately, most of the technologies needed to achieve the overhaul are already available and increasingly affordable. Many compare the shift we see in history from horses to cars or when aviation was taking off (pun intended). Imagine the amount of development it took to build roads, parking features, factories, and airports, not to mention all of the professionals that it took to create that change. But as people travel around the world, there are hardly any places that do not incorporate the same technology required for cars and planes.

Governments around the world announced ambitious climate pledges in 2020 when more than twelve countries plus the European Union proposed and even passed laws related to net-zero emissions. Governments also announced measures designed to adopt a range of clean energy technologies in response to the COVID-19 pandemic and the global energy crisis triggered by Russia's invasion of Ukraine. If fully implemented, currently announced manufacturing capacity expansions for solar PV and batteries would be sufficient to meet demand by 2030 according to the International Energy Agency.

## Mining for a New Industrial Revolution

The World Economic Forum's Great Reset of 2020 called for a new inclusive contract to rebuild the energy industry, to transform it into a resilient, equitable, and sustainable product to help reach the United Nations' "race to zero" (net-zero emissions) by 2050. While this chapter has discussed the challenges for top emitters like the United States, China, and India, to do that a controversial issue that has not been discussed is the concern for land degradation from the extraction of minerals needed for clean energy materials. Mining is not often seen in a positive light from the view of environmental activists. The international commitment to the Paris Agreement in 2016 and the required checkups or global stocktakes (a deadline every five years when nations present their accomplishments and their future plans for abiding to reduce greenhouse gas emissions) may not always highlight the methodology in their green energy transition. For energy demands to become sustainable, they need to be sourced from technologies that avoid the burning of fossil fuels, but green technologies carry intensive mineral demands. The new non-fossil fuel driven technologies will rely on non-renewable raw materials sourced from primary geological resources (mines) or secondary supply (reuse or recycling).

Richard Herrington, of Department of Earth Sciences and the Natural History Museum of London, UK, proposes a fully circular economy. Ideally, demand could be satisfied by reuse and recycling, but this practice is not yet at hand. There is not currently a secondary supply of materials, and recycling rates are still too small to meet demand. He reminds us in his 2021 paper that for materials necessary for green energy deployment such as aluminum and cobalt, of which up to 70 percent can be recycled, recycling accounts for only 30 percent of growing demand. Lithium recycling accounts for only 1 percent of present demand. In the time frame of reaching net-zero by 2050, it is not currently possible to use alternative technology solutions to reduce reliance on currently used minerals. Alternatives such as lithium-free multivalent metal-ion batteries to replace lithium-ion batteries, are less mature in their development and will take time to industrialize (Herrington 2021).

Sourcing challenges as well as infrastructure challenges leave us with mining as the necessary way to deliver ready-to-use technical solutions needed for the quickest road to decarbonization. Currently, over 1.4 billion internal combustion engine vehicles are on the roads globally, and the current movement toward wind, solar, wave, tidal, hydro, geothermal, and nuclear energy, along with the delivery methods currently used, are highly reliant on mineral-based technologies. Seventeen minerals are essential to clean energy in addition to cobalt and lithium. Depending on the type, photovoltaic cells require the mining of aluminum, copper, silver, iron, indium, selenium, and tellurium and the quarrying of silica and sand. Wind energy demands steel, copper, aluminum, zinc, and lead as well as neodymium for turbine magnets. Hydro power demands concrete and steel for basic infrastructure in addition to copper and aluminum for power transmission.

A small percentage increase in demand for any of these minerals will result in a large need for raw materials. For example, a 9 percent increase in aluminum by 2030 equates to 103 million tons of aluminum to be mined. To put that amount into context, it is more than the world's total annual production of aluminum in 2019, according to the 2020 World Bank Report. Another example is the amount of graphite needed, which is projected to be around sixty-eight million tons by 2050. Cobalt, an essential mineral for green energy, is predominantly mined in the Republic of the Congo, where political instability could halt a steady supply. Deep ocean nodules could offer ready supplies of cobalt, manganese, copper, and nickel needed for the world's battery electric vehicles, but mining the ocean floor is difficult, expensive, and controversial.

How can the traditionally invasive mining industry be made sustainable? The mining sector's carbon emissions make up an estimated 28 percent of global emissions, and reducing these numbers will require drastic measures. According to the IEA 2021 Net Zero by 2050 Report, the mining of minerals will need to grow six times by 2050 to achieve global net zero carbon emissions. Sustainable mining would be a clear answer in the development of mining infrastructure going forward, working toward a circular economy where extraction becomes minimal in the supply chain and international recycling is promoted. ExxonMobil declared on their web page that mining is going through "seismic changes" to meet their sustainability goals. They and other mining/energy giants know that they are facing increased environmental scrutiny. Many mining companies have committed to modernizing their infrastructure by electrifying machinery and vehicles, by encouraging research and innovation development, and by using artificial intelligence (AI).

Large mining corporations note that AI has already had a major impact on the mining industry, including operations. AI analyzes geological data patterns and historical mining data to provide accurate mineral reserve estimates. This helps mining companies optimize resource utilization and plan operations effectively while minimizing risk and environmental damage. By converting vast unstructured data into actionable insights, AI identifies patterns and anomalies, allowing mining companies to discover untapped opportunities and make informed decisions proactively. AI-driven, real-time monitoring promotes predictive maintenance, improving safety and efficiency and reducing downtime. AI technologies have already played key roles, including computer vision, smart robots, data, and machine learning. AI reshapes mining by combining data-driven insights with innovative technologies, driving efficiency, safety, and sustainability.

## Solutions

To mitigate means to make less severe, less painful, or to lessen the gravity of an offense or mistake (Merriam Webster 2011). Mitigating or reducing the effects of climate change involves strategies that reduce the amount of heat trapped in the atmosphere so that significant human-caused changes to the Earth's climate are avoided. The 2014 United Nations Intergovernmental Panel Report on Mitigation stated that the goals of mitigation efforts are to stabilize greenhouse gas levels within time to allow ecosystems to adapt naturally to climate change, and to ensure that food production is not imperiled, and finally to let economic development proceed in a sustainable manner. As discussed previously, many governments and corporations are committed to reducing the amount of greenhouse gases emitted but scientists worry that the reduction will not be fast enough. The 6th Assessment Report published by the IPCC in 2023 urged that removal of greenhouse gases from the atmosphere must be included if the Paris Agreement goals are to be met. Scientists and policy makers are looking at combined strategies to decrease greenhouse gases as soon as possible. This section looks at the various ways, including physical and social, that could help the international community reach Paris Agreement goals.

## Reducing Greenhouse Gases in the Atmosphere

Perhaps the most obvious method for achieving the goals of the Paris Agreement is to reduce the amount of human-generated emissions of greenhouse gases. After all, that is the cause of the problem. Since 2020, a little over 73 percent of global greenhouse gas emissions or $CO_2$ equivalent (greenhouse gases standardized to the global warming potential of one ton of $CO_2$) have come from the energy sector, followed by agriculture, forestry, and land-use at approximately 18 percent, industry (chemical manufacturing and cement) at 5.2 percent, and waste at 3.2 percent (Friedrich et al. 2023).

Therefore, a reduction in the emission by the largest contributor, the energy sector, would logically make quite an impact. The current efforts to combat emissions in the energy sector largely rely on the use of carbon capture and storage (CCS) or carbon capture utilization and storage (CCUS) as a transitional method for the continued use of fossil fuels. As mentioned earlier in this chapter, CCUS refers to a variety of techniques that are used to trap the $CO_2$ emitted by power plants and other industrial facilities and the use of the trapped $CO_2$ for other purposes. The remaining $CO_2$ is sequestered or stored where it cannot escape. The United States and other countries are currently spending billions of dollars to develop storage facilities, but many argue that the money could be better spent in the development of renewable energy sources. The counter argument is that renewable energy sources are not yet in place to provide enough energy to large populations and battery storage properties are not yet viable to store energy from renewables when wind or solar is not available. The hope still remains that the development and deployment of clean energy such as green hydrogen, wind, solar, nuclear, and biofuels will eventually be the main sources of energy.

## Enhancing Carbon Sinks

Nature has been capturing $CO_2$ for eons. Natural climate solutions have a key role to play in the race to net-zero. The World Economic Forum estimated that natural carbon sinks, environments that capture and store carbon, already absorb half of the human-produced carbon emissions. In the simplest of terms, a carbon sink is anything that absorbs more carbon than it produces. This includes natural environments that can store captured carbon indefinitely. These natural systems are vulnerable, however, and their destruction can release stored carbon back into the atmosphere.

### *Forests*

NASA satellite data shows the world's forests absorb 15.6 billion tons of $CO_2$ a year. Burning or eliminating a forest is a compounding issue. The loss of the trees reduces the size and capability to draw-down $CO_2$ or "sink" it, and the burning or decomposition of the tree, releases greenhouse gases into the atmosphere. Wildfires and deforestation release almost half of their sequestered $CO_2$ (as much as one gigaton) back into the atmosphere, a fact that stresses the need to protect forests so they don't contribute to the anthropogenic emissions of forty gigatons per year or 2,400 gigatons since the Industrial Revolution. Planting trees is often the "go-to offer" by companies or governments when it comes to their pledges of helping reach net-zero goals. Trees are a ready-to-go kind of negative emissions "technology," and many see them as an easy, proven, natural way to remove carbon from the atmosphere. In its January 2020 report, "CarbonShot," the World Resources Institute, a global nonprofit research organization, suggested large and immediate investments in reforestation within the United States to reduce carbon emissions and enhance carbon withdrawal to reach net-zero by 2050. They urged the US government to invest $4 billion a year through 2030 to support tree restoration projects nationally.

Forestry and agriculturalist specialists, like Lalisa Duguma of World Agroforestry in Nairobi, Kenya, emphasize that the movement of tree planting needs to shift from planting to growing. In 2020 she published a paper looking at success rates of large planting operations. She noted that if there is hope of success, restoration projects need to consider some very important factors: (1) the best times of year to plant seeds, (2) which seeds to plant, (3) the best location for the seeds, (4) who will care for the seedlings as they grow into trees, (5) how that growth will be monitored, and (6) how to balance the economic and environmental needs of people in developing countries where the trees might be planted. This approach may reduce the likelihood that these projects fail. A disheartening example of mismanagement was the efforts in 2019 in Turkey's Çorum province on National Forestation Day. Volunteers across the country planted eleven million trees at more than 2,000 sites. Within three months, though, up to 90 percent of the new saplings inspected by Turkey's agriculture and forestry trade union were dead. The saplings died due to a combination of insufficient water and because they were planted at the wrong time of year and not by experts. Planting a monoculture of trees can also backfire as it reduces biodiversity. The key take-away is that planting a forest is not the simple answer many hope it will be. Efforts like China's Green Wall in the Gobi Desert, where trees were planted to hold back encroaching sand dunes and to increase a natural carbon sink, are not often successful unless trees are seen as part of particular ecosystems and not like home air filters that can be put anywhere. The green wall of the Sahel in Africa suffered similar consequences (Duguma et al. 2020).

Trees also play a large part in weather events through their involvement in the hydrologic cycle. New National Science Foundation supported research published in the National Academy of Sciences led by University of Utah biologists William Anderegg, Anna Trugman, and David Bowling found that some plants and trees may actually promote drought conditions. As trees siphon water out of the soil drought conditions intensify. Anderegg and colleagues report that because trees take up, transport, and evaporate water, they can influence societally important extreme events, like severe droughts. Plants and forests draw water out of the soil and exhale it into the atmosphere, a process known as transpiration. This impacts the balance of water and heat at our planet's surface, which fundamentally controls the weather. In some cases, such as in the Amazon rainforest, water vapor can jumpstart precipitation, but in a dry region, planting the wrong type of water-thirsty tree could increase drought conditions (Anderegg et al. 2019).

The location where trees are planted must also be considered, especially because of extreme weather events such as heatwaves. Tree planting efforts must consider that climate change adds a myriad of threats to trees. For example, in places like California, extreme storms, droughts, and resulting disease and insect infestations are stressing and killing trees. The dead trees become fuel for wildfires. The Dixie Fire in Northern California, which burned more than 950,000 acres, was likely sparked by a tree falling onto a power line. University of California Berkeley researchers study tree mortality, climate change, and their struggle with increased heat. Many species are unable to tolerate increasing temperatures and instead of aiding in drawing down $CO_2$, they have high mortality rates and add to the problem of carbon being released through wildfires.

### *Oceans*

People often think of forests as the "lungs" of the planet, but the United Nations clarifies on their web page that it is in fact the ocean that should have this distinction. The oceans produce a significant amount of Earth's oxygen—about half of the oxygen we breathe comes from marine sources. The title of "lungs" of the planet is supported by the IPCC Special Report on the Ocean and Cryosphere in a Changing Climate where it states that the ocean has absorbed more than 90 percent of the excess heat in the climate system and since the 1980s has absorbed 20 to 30 percent of anthropogenic $CO_2$ emissions. It is also the Earth's most important natural climate moderator, regulating climate and influencing weather patterns around the globe and affecting every location on Earth. Despite this moderation, changes have still been felt around the world with warmer seas linked to larger, more intense atmospheric rivers and hurricanes in the Pacific and Atlantic, increasing droughts in Central Africa, changing precipitation patterns in the midwestern United States, and more intense flooding in South-East Asia as mentioned in Chapter 1.

Despite the toll that human-produced greenhouse emissions have had on the oceans (including loss of biodiversity, acidification, and coral bleaching), the Ocean Conservancy points out that the ocean is much more than a victim of climate change. It is also a critical part of the solution to it. A report commissioned by the High-Level Panel for a Sustainable Ocean Economy found that by 2050 ocean solutions could contribute as much as one-fifth of the annual greenhouse gas (GHG) emission cuts needed to limit global temperature rise to 1.5 °C. At the same time, ocean-based solutions can create jobs and restore ocean health. Ocean-based climate solutions reduce GHG emissions and increase the resilience of people and nature in the face of climate impacts already underway. They are unique in that they are designed for ocean and coastal communities. They also consider the international aspect of ocean governance (Wood and Ashford 2023).

As mentioned earlier, energy is the global sector most responsible for greenhouse gas emissions. Developments in ocean-based clean energy include offshore wind to wave and tidal energy. The United Kingdom is the world's largest offshore wind energy producer and has committed to generating over 30 percent of its electricity from offshore wind by 2030. The United States has committed to powering more than ten million homes with offshore wind, enough to cut seventy-eight million tons of carbon dioxide ($CO_2$). Offshore wind, in addition to floating solar and tidal power, could cut greenhouse gas emissions by up to 3.60 gigatons per year in 2050 (an amount larger than the total combined emissions for the entire European Union in 2021). Increasing ocean-based renewables will also be critical to meeting global energy demand as the world works to phase out coal and other fossil-based energy sources.

Global pledges in offshore wind have approximately doubled in the last four years, bringing targeted deployment to up to 2,000 GW (enough to power approximately 1.5 billion homes annually) by 2050. Certain countries are leading by example: Norway, for instance, is home to the world's largest fully operational floating offshore wind park, Hywind Tampen. It has also allocated areas for 30 GW of offshore wind power production by 2040 for a country whose average household consumes about 7,000 kWh. If countries are committed to well-planned and responsibly developed ocean-based renewable energy, it could provide one-tenth of global electricity demand and save over three billion tons of $CO_2$ annually. Development, however, must be done responsibly with consideration to ocean ecosystem, migration routes, and interference with currents important to wildlife.

Blue carbon is the term for carbon captured by the world's ocean and coastal ecosystems. It includes mangrove forests, tidal marshes, and seagrass meadows that are gaining international recognition as a natural climate solution currently ready and able to contribute to climate change mitigation and adaptation targets. These ecosystems can sequester three to five times more carbon per area than tropical forests and absorb it from the atmosphere around three times as quickly when they are at their healthiest. These essential ecosystems are also important buffers from storms and retention areas for floodwaters. The IPCC has adopted ways in which countries can include the protection of wetlands as a way that they are committed to reducing emissions and protecting ecosystems, biodiversity, and communities from extreme weather. Recognizing this, as part of its updated nationally determined contribution to the Paris Agreement, Costa Rica pledged to protect 100 percent of its coastal wetlands by 2025. The protection of wetlands also hits the nail on the head as they purify water and reduce erosion. According to Ocean Conservancy, if the 151 countries that have at least one blue carbon ecosystem do the same, emissions reductions could roughly equal all of South

Korea's emissions. Protection and restoration of existing blue carbon ecosystems could avoid emissions of approximately 3 percent (0.5–0.8 percent from protection and 2.3–2.5 percent from restoration) of annual global greenhouse gas emissions.

The continued development and augmentation of seafood-based food production is also a scalable solution. The increasing global population will require additional protein sources. The ocean can play a key role in meeting this need. Algae, fish, and invertebrates like shellfish require fewer resources to produce than options like beef and lamb. Incorporating these "blue" foods into global diets not only diversifies protein choices but could also reduce global emissions by up to 1.47 gigatons per year in 2050, comparable to removing 393 coal-fired power plants annually (Gephart et. al. 2021). Governments must promote their production and subsidize their consumption as currently they are a more expensive protein source.

One of the most challenging ocean-based solutions involves harvesting carbon from the atmosphere through direct air capture or waste combustion and pumping the liquefied carbon into chambers below the seabed, where it can be stored permanently. This technique is discussed in more detail in the Climate Engineering section of this chapter. Another approach involves removing $CO_2$ from the ocean. Marine $CO_2$ removal includes a range of techniques, including ocean alkalinity enhancement, in which alkaline minerals, like olivine, or artificial substances, like lime or industrial byproducts, are added to seawater to enhance the ocean's natural carbon sink (removal of $CO_2$ from the atmosphere). It involves a series of reactions that convert dissolved $CO_2$ into stable bicarbonate and carbonate molecules. This in turn causes the ocean to absorb more $CO_2$ from the air to restore equilibrium. The finely ground alkaline substances can be spread over the open ocean or deposited in alkaline sand or gravel on beaches or coastal sea beds. Engineers have proposed reacting seawater with alkaline minerals inside specialized fuel cells before releasing it back into the ocean. In this method, alkaline minerals are added into the ocean to alter its chemistry and increase carbon uptake. Ocean nutrient fertilization produces massive algal blooms that would take up huge amounts of carbon. Theoretically these practices appear feasible but have not advanced to a trial stage. Ocean floor carbon storage and marine $CO_2$ removal must undergo ecological and socioeconomic impact assessments, and policy and governance questions must be addressed.

In addition to its potential for reducing greenhouse gas emissions, the ocean has a key role to play in ensuring all life on Earth is resilient in the face of extreme weather and other climate change impacts. The Ocean Conservancy points out that building climate resilience needs to include the management of fisheries for the impacts of climate change, ensuring that critical marine areas are protected with climate impacts in mind. Policy needs to include the understanding of new migration patterns and the protection of species as they move inland or to cooler waters. Incidentally, according to the Ocean Conservancy, of the forty coastal countries that submitted or updated their Nationally Determined Contributions in 2023, thirty-seven of them included ocean-based climate solutions. Costa Rica, Fiji, Chile, and Cabo Verde are among those leading the way. These countries included ambitious commitments to invest in blue carbon, decarbonize shipping, and expand protected areas.

New research commissioned by the High-Level Panel for a Sustainable Ocean Economy reported that ocean-based climate solutions can deliver up to 35 percent of the annual greenhouse gas emission cuts needed in 2050 to limit global temperature rise to 1.5 °C (2.7 °F). As the World Economic Forum's 2023 report on the Voluntary Carbon Market pointed out, oceans, forests and other natural carbon sinks are some of the few solutions that are ready today for carbon removal. They and hundreds of other scientists stress the importance of protection and proliferation of natural carbon sinks.

## Social Science Solutions

The nations and global corporations that are moving forward to meet the goals of the Paris Agreement are often said to be using a method that motivates and addresses climate solutions under a social lens. That method is *climatization*. It encompasses the idea that major decisions are being made under the lens of climate change mitigation and resilience. The public can see that climate topics dominate global discussions by the G7, the intergovernmental political and economic group with members whose governments share political values of pluralism, democracy, and representative government (currently Canada, France, Germany, Italy, Japan, the United Kingdom, the United States, and the European Union). As leaders consider mitigation and adaptation strategies that may lessen the risk of extreme weather events, they may move toward small incremental changes or large societal shifts. Of course, any effort has overlapping economic and social considerations. Because we are living within a new climate regime, experts of this topic, such as Stefan Aykut (a fellow at the Humanities Centre for Advanced Studies in Hamburg, Germany) and Lucile Maertens (a faculty member of Social and Political Sciences, Institute of Political Studies, Lausanne, Switzerland), think that climatization is here to stay. While climatization may be the new practice for leaders making an ethical decision on the future of their institutional framework, they share common challenges.

Climate litigation is also a key way people are using societal structures in the wake of extreme weather disasters. According to the United Nations, climate litigation has more than doubled from 2018 to 2023. A report by the Sabin Center for Climate Change Law at Columbia University shows that climate litigation is becoming an integral part of securing climate action and justice. As climate litigation increases in frequency and volume, the body of legal precedent grows, forming an increasingly well-defined field of law. The report, "Global Climate Litigation Report: 2023 Status Review," documents focused on climate change law, policy, or science collected by the Sabin Center's US and Global Climate Change Litigation Databases (Burger and Tigre 2023). While most cases were held in the United States, climate litigation is taking root all over the world, with about 17 percent of cases happening in developing countries, including Small Island Developing States. Worldwide, the cases are hosted at international, regional, and national courts, tribunals, quasi-judicial bodies, and other adjudicatory bodies, including special procedures of the UN and arbitration tribunals.

According to the 2023 report, most ongoing climate litigation falls into one or more of six categories: (1) cases relying on human rights given by international law and national constitutions; (2) the lack of enforcement of climate-related laws and policies; (3) those advocating to keep fossil fuels in the ground; (4) advocates for greater climate disclosures and an end to greenwashing; (5) claims addressing corporate liability and responsibility for climate harms; and (6) claims bringing up failures to adapt to the impacts of climate change. The report demonstrates how courts are finding strong human rights linkages to climate change. This is leading to greater protection for the most vulnerable groups in society. Outcomes of the cases have also increased accountability, transparency, and justice, encouraging governments and corporations to pursue more ambitious climate change mitigation and adaptation goals. The report also predicts a rise in the number of cases that will be filed due to climate migration, cases brought by Indigenous peoples, local communities, and other groups disproportionately affected by climate change. They also foresee a rise in cases addressing liability following extreme weather events.

## Climate Engineering

Because keeping the Earth's temperatures from rising is dependent on balancing the amount of solar radiation received, withheld by the atmosphere, and released back into space, many climate change solutions incorporate the manipulation of one of those functions. Several methods have been proposed to stabilize the Earth's energy budget. This section focuses on efforts to mitigate climate change and extreme weather events by first and foremost reducing greenhouse gases in the atmosphere, increasing the amount of radiation that exits back out to space, and decreasing the amount of solar radiation that hits the surface of the Earth. Scientists, engineers, and entrepreneurs have two choices: to enhance or mimic biological processes at a rate fast enough to keep global temperatures from increasing above 1.5 °C or to try and accomplish the same with human-made engineering efforts (geoengineering). Some engineering feats are already being employed and have proven somewhat successful even though they are not at the scale needed. Other proposals sound like they belong in science fiction but are actually feasible in theory. With each and every idea, the public raises concerns about cost, efficiency, reliability, and complications or damage to other processes of nature.

### *Alternative Fuel Sources*

When it comes to the specific goal of reducing greenhouse gases from fossil fuels, the answer is of course to find an alternative fuel source with lower emissions. Alternative fuels are attractive to governments and corporations, as they can reduce their contribution to smog, air pollution, and global warming. Most alternative fuels are also sustainable, not finite, and most countries have the materials needed to produce them, offering energy independence.

The US Energy Policy Act (EPAct) of 1992 identified eight fuels as having the potential to serve as full or partial alternatives to gasoline or diesel. They include methanol, ethanol, and other alcohols; blends of 85 percent or more of alcohol with gasoline; liquid fuels domestically produced from natural gas as well as natural gas; propane; hydrogen; electricity; biodiesel; coal-derived liquid fuels; fuels, other than alcohol, and fuels derived from biological materials. The Department of Energy has the authority to add more alternative fuels if certain criteria are met. Most promising are hydrogen fuels, especially blue and green hydrogen, as we discuss next.

Currently, we do not have a large enough supply of emissions-free alternative fuel sources to power industries that use high density energy vehicles and machinery. These include the long-distance transportation industry, those that require intense heat production (such as steel manufacturing), the chemical production industry, and the cement production industry. Low-carbon fuel sources like hydrogen may provide the energy required. Hydrogen is a ubiquitous (found everywhere) element. However, 99 percent of hydrogen in industry today is produced from hydrocarbons, namely natural gas (methane) and coal. As a result, its production emits carbon dioxide ($CO_2$). This type of hydrogen is often referred to as blue hydrogen produced using a process called "steam reforming," whereby steam separates hydrogen from natural gas. The process emits GHGs, but CCS technologies capture and store those emissions. Methane ($CH_4$) is split into hydrogen and $CO_2$. In 2017, global hydrogen production accounted for more annual $CO_2$ emissions than both the entire nation of Germany and the global shipping industry. CCS has allowed fossil fuel companies to evolve and strive to meet emissions reductions goals. While hydrogen has a huge potential to decarbonize heavy-emitting sectors like those involved in making steel, cement, fuels, and petrochemicals, without carbon storage (a technology that may not be available in developing nations), the production of blue hydrogen emits immense amounts of greenhouse gases. Green hydrogen however, is produced by wind and solar via electrolysis, or the splitting of water molecules into hydrogen and oxygen atoms. It is produced without the use of hydrocarbons and can greatly reduce a nation's or company's emissions.

### *Carbon Capture and Storage*

The process of CCS was briefly discussed in our section on India. It involves a range of technologies that either stop $CO_2$ from entering the atmosphere, by "filtering out" $CO_2$ on the way to the smokestack of a facility such as a power plant or factory, or by removing $CO_2$ that's already in the air, a process known as "direct air capture." Direct air capture is more expensive because air typically contains only about 0.04 percent of the gas. In both cases, the $CO_2$-containing gases or air are typically absorbed by a solvent or solution, and then separated out again later. So far, a majority of CCS projects around the world focus on the first. Once the carbon is captured, it can either be permanently stored underground or transformed into a carbon-containing product referred to as carbon capture and utilization or CCU. In Canada for example, most projects involve CCU, and the captured $CO_2$ is then used to push more oil out of aging oil wells and then stored underground in air pockets in the porous rock of depleted wells, with a "cap stone" on top to prevent leakage.

In the United States, Republicans, Democrats, and Independents agree on carbon capture. The Lawrence Livermore National Lab (LLNL) has been at the forefront of innovations involving this solution in recent years. The lab was originally established in 1952 to meet national security needs through engineering and technological advances. At its inception, the LLNL was primarily tasked with the nuclear weapon industry following the Cold War. Now, they have expanded to include research in the sectors of counterterrorism, intelligence, defense, and energy. A challenge for the lab is to devise ways to enhance national energy security without increasing the impact that energy production has on the planet. LLNL scientists have especially focused on removing existing greenhouse gases from the atmosphere with negative emissions technologies.

Encouraged by tax incentives included in Biden's Inflation Reduction Act, some companies in the United States have developed CCS facilities and plan to increase that technology. Although this looks to be a promising endeavor, some worry that CCS could distract from efforts to develop cleaner energy and reduce emissions industrywide. CCS is one of the few climate change solutions whose goal is not to reduce the amount of $CO_2$ emitted but to actually remove $CO_2$ from the atmosphere, thereby generating "negative emissions." Many politicians and scientists think this may be essential if we are to limit global warming to 1.5 °C above pre-industrial times—the more stringent of the two targets under the Paris Agreement that aims to prevent the worst impacts of climate change including extreme weather. Last year's special report from the Intergovernmental Panel on Climate Change found $CO_2$ removal was needed in every successful scenario to cancel out emissions from sources for which no mitigation measures have been identified such as long-distance air travel and cement production.

### *Theoretical Geoengineering*

Scientists are working on geoengineered methods to secure a possible way to avoid going beyond the tipping point when chances of extreme storms, drought, heat waves, and wildfires would increase significantly, and humans and other species would struggle to survive. Theoretical geoengineering projects are underway in many of the world's top universities and physics/atmospheric laboratories. While many of these projects seem like incredibly far-off imaginative solutions, the scientists working on them are in the mind-frame that these methods may be needed if temperatures continue to rise. However, theoretical solutions to global warming often require an extreme amount of interference of the sun's rays or changing the physics or chemistry of the planet. Climate scientists that model both a pre-industrial Earth's climate and a post geoengineered climate have shown that a large difference exists between the two. Some climate modelers testing geoengineering methods urge that engineering efforts should stick to the development of clean energy and carbon capture instead of ways to offset the fourfold amount of emissions added to the atmosphere in the last 200 years.

If we think about heat in the atmosphere, whether it is a consequence of the natural greenhouse effect or the anthropogenic greenhouse effect, it all stems from the sun. The sun emits an enormous amount of energy toward the Earth. Luckily, some of that energy is reflected back out to space by a process known as the albedo effect, or the fact that some materials, such as clouds and other reflective surfaces like snow, ice, and lighter colored materials reflect the rays back out to space. The sun's rays that are not reflected back out to space are absorbed by darker colored surfaces, such as black tops and the dark blue ocean. What if we could lessen the amount of the sun's energy that reaches the surface of the Earth in the first place?

One strategy mimics the cooling effect that happens when large volcanoes erupt and emit large amounts of ash and other particulate matter that remains suspended in the air circulating in the atmosphere. An example of the influence of volcanic eruptions causing extreme weather shifts is the infamous "year without a summer." In April of 1816, the largest recorded eruption in 2,300 years occurred. The massive eruption of Mount Tambora in the Dutch East Indies (modern-day Indonesia) caused the average global temperatures to decrease by 0.4 to 0.7 °C (0.7–1 °F). Summer temperatures in Europe that year were the coldest ever from data from 1766 to 2000. The results were massive crop failures and famine across the Northern Hemisphere. In 1991, Mount Pinatubo erupted in the Philippines, projecting twenty million tons of sulfur dioxide aerosols into the stratosphere. The US Geological Survey reported that the Earth's lower atmosphere temperature dropped by approximately 0.7 °C or 1 °F. The effect only lasted a couple of years. If the temperatures continue to climb, could humans use the same principles of blocking the sun with aerosols to cool global temperatures?

Scientists propose that in much the same way that volcanic aerosols block and reflect incoming solar radiation, so too can other aerosols be manually injected into the lower stratosphere (approximately 20 km up). The idea was published in a peer-reviewed journal (Smith and Wagner 2018) along with the costs associated in using this method within the next fifteen years. The solution would entail pumping sulfates or other aerosol particles into the atmosphere that would spread around the planet and reflect sunlight back to space. The authors calculated an average cost of approximately $2.25 billion a year over the first fifteen years of deployment. The proposed number of flights were listed at approximately 4,000 in the first year. While no current design of airplane could facilitate the endeavor, the authors concluded that designing a workable airplane would not be difficult or expensive. Climate and weather models run by scientists at the National Academy of Science convention showed that this idea might also weaken Atlantic hurricanes. Unfortunately, the models also show that this method might have severe risks to global climate, particularly increasing drought in Africa's Sahel region, and may disrupt the Asian monsoon that so many rely on for agriculture. An interesting dilemma with this plan is that if for whatever reason the aerosol spraying stopped, temperatures would climb at breakneck pace within a few years—too fast for many biological systems to adjust to, including humans.

Adding aerosols to the lower stratosphere is not the only proposal out there to shade the Earth from solar radiation. A recent study led by the University of Utah explored scattering dust deep into space, and at the Massachusetts Institute of Technology, scientists are looking into creating a shield made of "space bubbles." Researchers at the University of Hawaii published a paper that suggested tethering a big solar shield to a repurposed asteroid. Most recently, scientists lead by Yoram Rozen, a physics professor and the director of the Asher Space Research Institute at Technion-Israel Institute of Technology, are working on a huge sunshade to be theoretically suspended between the Earth and the sun to block a small but crucial amount of solar radiation, which they think may be enough to block a little less than 2 percent of the sun's radiation and cool the planet by 1.5 °C (2.7 °F). Rozen's team is not the only one working on sun shades. Similar versions of this approach are gaining popularity as temperatures continue to climb. Increasing the albedo (or reflectance) of clouds is another proposed solution to decrease the amount of absorbed solar radiation. Meteorologist Alan Gadian, senior scientist at the UK's National Centre for Atmospheric Science, is among a group pursuing research to cool strategic parts of the warming sea, such as the Gulf of Mexico and the Atlantic's hurricane alley between Africa and the Caribbean, by "marine cloud brightening." The method was first proposed in 1990 by cloud physicist John Latham at the National Center for Atmospheric Research in Boulder, Colorado.

The idea is to infuse clouds with particles of sea salt, around which water vapor would condense to form droplets. The more droplets in a cloud, the whiter it is, and the more sunlight it reflects, cooling the sea below it. The theory is that marine cloud brightening could be activated prior to hurricane season to cool ocean hot zones and prevent superstorms. They envision a fleet of ships drawing up seawater and spraying stratocumulus clouds overhead with a constant mist. If the spraying started in November in the Atlantic south, ocean gyres (circulation systems of the ocean) would carry the cooled water north into the hurricane producing regions in time for the next hurricane season.

A similar proposal to augment the albedo effect involves the reflective properties of ocean water. A project coined "Bright Water" proposed by physicist Russell Seitz would make the sea itself more reflective, rather than the clouds above it. Ships would pump microscopic bubbles into their wake along tropical storm tracks, cooling the water. Seitz theorizes that this method would have a lower risk of affecting the weather than pumping aerosols into clouds. While some scientists hope that engineering efforts would re-balance the Earth's energy, the American Meteorological Society submitted an official policy statement regarding geoengineering, warning that "Reflecting sunlight would likely reduce Earth's average temperature but could also change global circulation patterns with potentially serious consequences such as changing storm tracks and precipitation patterns" (American Meteorological Society 2009).

Yet another way to lower ocean temperatures would be to mimic the process of upwelling where deep ocean water is brought to the surface. Upwelling through geoengineering could be done by stirring ocean waters, mixing warm surface water into the cold depths and bringing cooler water to the surface. A theoretical proposal to do this would be to have an array of long vertical tubes, hanging like drainpipes from enormous floating rings, reach down into the cold water. As waves lap over the rings, the water level inside them rises above the surrounding sea. This would push the whole column of warm water downward and force cooler water up to the surface.

## Further Reading

Abdi, H. 2017. "Where Is the Most Vulnerable Country to Climate Change?" Climate Home News, June 13, 2017. Accessed October 14, 2024. https://www.climatechangenews.com/2017/06/13/vulnerable-country-climate-change.

Alleyne, B. 2019. "Mia's Climate Plea at UN." *Nation News*, September 28, 2019.

American Meteorological Society. 2009. "Geoengineering the Climate System 2009." Accessed October 14, 2024. https://www.ametsoc.org/index.cfm/aMS/about-ams/ams-statements/archive-statements-of-the-ams/geoengineering-the-climate-system-2009.

Anderegg, W. R. L., A. T. Trugman, G. Badgley, A. T. Trugman, G. Badgley, C. M. Anderson, A. Bartuska, P. Ciais, D. Cullenward, C. B. Field, J. Freeman, S. J. Goetz, J. A. Hicke, D. Huntzinger, R. B. Jackson, J. Nickerson, S. Pacala, and J. T. Randerson. 2019. "Climate-driven Risks to the Climate Mitigation Potential of Forests." *Science* 366.6463: eaaz7005.

Angel, M. 2021. "Protecting and Enhancing Natural Carbon Sinks: Natural Climate and Community Solutions." MIT Climate Portal, May 4, 2021. Accessed October 14, 2024. https://climate.mit.edu/posts/protecting-and-enhancing-natural-carbon-sinks-natural-climate-and-community-solutions.

Asia Society Policy Institute. 2022. "Getting India to Net Zero." August 26, 2022. Accessed October 14, 2024. https://asiasociety.org/policy-institute/getting-asia-net-zero/country-report-india.

Bailey, A. J., C. M. Wills, and J. Mitchem. 2022. "Attitudes Towards Climate Change and Scientific Stories." *Journal of Environmental Studies and Sciences* 12: 714–26. https://doi.org/10.1007/s13412-022-00770-0.

Bradsher, K. 2023. "Why Heat Waves Are Deepening China's Addiction to Coal." *The New York Times*, July 20, 2023. Accessed October 14, 2024. https://www.nytimes.com/2023/07/20/business/china-coal-climate-change.html.

Burger, M., and M. A. Tigre. 2023. "Global Climate Litigation Report: 2023 Status Review." Sabin Center for Climate Change Law, Columbia Law School & United Nations Environment Programme.

Carman, J., S. Rosenthal, E. Maibach, J. Kotcher, X. Wang, J. Marlon, and A. Leiserowitzet. 2020. "Shifting Views of an Ideal Society in the U.S." Yale Program on Climate Change Communication, November 17, 2020. Accessed October 14, 2024. https://climatecommunication.yale.edu/publications/shifting-views-of-an-ideal-society-in-the-u-s.

Childress, S. 2012. "Timeline: The Politics of Climate Change." PBS, October 23, 2012. Accessed October 14, 2024. https://www.pbs.org/wgbh/frontline/article/timeline-the-politics-of-climate-change.

Climate Bonds Initiative and China Central Depository & Clearing Research Centre (CCDC Research). 2021. "China Green Bond Market Report 2021." Accessed October 14, 2024. https://www.climatebonds.net/files/reports/cbi_china_sotm_2021_0.pdf.

CNN. 2019. "I Used to Be a Climate Skeptic. This Fact Changed My Mind." September 4, 2019. Accessed October 14, 2024. https://www.cnn.com/videos/weather/2019/09/03/climate-change-global-warming-skeptic-chad-myers-nws-orig.cnn.

Cui, R. Y., N. Hultman, D. Cui, H. McJeon, S. Yu, M. R. Edwards, A. Sen, K. Song, C. Bowman, L. Clarke, J. Kang, J. Lou, F. Yang, J. Yuan, W. Zhang, and M. Zhu. 2021. "A Plant-by-Plant Strategy for High-Ambition Coal Power Phaseout in China." *Nature Communications* 12, no. 1 (March 16). https://doi.org/10.1038/s41467-021-21786-0.

Dang, H.-A., S. Hallegatte, and T.-A. Trinh. 2023. "Does Global Warming Worsen Poverty and Inequality? An Updated Review." SSRN, November 4, 2023. Accessed October 14, 2024. https://papers.ssrn.com/sol3/papers.cfm?abstract_id=4622015.

Denchak, M. 2021. "Paris Climate Agreement: Everything You Need to Know." Be a Force for the Future, February 19, 2021. Accessed October 14, 2024. https://www.nrdc.org/stories/paris-climate-agreement-everything-you-need-know.

Duguma, L., P. Minang, E. Aynekulu, S. Carsan, J. Nzyoka, A. Bah, and R. Jamnadass. 2020. "From Tree Planting to Tree Growing: Rethinking Ecosystem Restoration Through Trees." Working Paper. https://doi.org/10.5716/wp20001.pdf.

English, B. J. 2021. "Maslow's Hierarchy of Needs and the Psychology of Climate Change." Accessed August 1, 2024. https://core.ac.uk/download/524536204.pdf.

EPA. 2021. "EPA Report Shows Disproportionate Impacts of Climate Change on Socially Vulnerable Populations in the United States." September 2, 2021. Accessed October 14, 2024. https://www.epa.gov/newsreleases/epa-report-shows-disproportionate-impacts-climate-change-socially-vulnerable.

Eyring, V., N. P. Gillett, K. M. Achuta Rao, R. Barimalala, M. Barreiro, N. Bellouin, C. Cassou, P. J. Durack, Y. Kosaka, S. McGregor, S. Min, O. Morgenstern, and Y. Sun. 2021. "Human Influence on the Climate System." In *Climate Change 2021: The Physical Science Basis. Contribution of Working Group I to the Sixth Assessment Report of the Intergovernmental Panel on Climate Change*, Masson-Delmotte, V., P. Zhai, A. Pirani, S. L. Connors, C. Péan, S. Berger, N. Caud, Y. Chen, L. Goldfarb, M. I. Gomis, M. Huang, K. Leitzell, E. Lonnoy, J. B. R. Matthews, T. K. Maycock, T. Waterfield, O. Yelekçi, R. Yu, and B. Zhou (eds.) Cambridge, UK: Cambridge University Press, pp. 423–552, https://doi.org/10.1017/9781009157896.005.

Frank-Oster, C. 2013. "A Perfect Moral Storm: The Ethical Tragedy of Climate Change." *Ethics, Policy & Environment* 16, no. 3 (October): 356–65. https://doi.org/10.1080/21550085.2013.844579.

Friedrich, J., M. Ge, A. Pickens, and L. Vigna. 2023. "This Interactive Chart Shows Changes in the World's Top 10 Emitters." World Resources Institute, March 2, 2023. Accessed October 14, 2024. https://www.wri.org/insights/interactive-chart-shows-changes-worlds-top-10-emitters.

Gan, N. 2022. "Record Rainfall and Scorching Heat Wave Cause Havoc in China." CNN, July 21, 2022. Accessed October 14, 2024. https://www.cnn.com/2022/07/20/china/china-summer-extreme-weather-climate-change-intl-hnk-mic/index.html.

Gardiner, S. M., and L. Hartzell-Nichols. 2012. "Ethics and Global Climate Change." *Nature Education Knowledge* 3, no. 10: 5.

Geostrata, The. 2022. "Cognitive Dissonance—Climate Change." October 31, 2022. Accessed October 14, 2024. https://www.thegeostrata.com/post/cognitive-dissonance-climate-change.

Gephart, J. A., Patrik J. G. Henriksson, Robert W. R. Parker, A. Shepon, K. D. Gorospe, K. Bergman, G. Eshel, C. D. Golden, B. S. Halpern, S. Hornborg, M. Jonell, M. Metian, K. Mifflin, R. Newton, P. Tyedmers, W. Zhang, F. Ziegler, M. Troell. 2021. "Environmental Performance of Blue Foods." *Nature* 597: 360–5.

Global Voices. 2023. "Is India's Transition to Clean Energy Achievable or an Empty Promise?" #EthicalHour, April 6, 2023. Accessed October 14, 2024. https://ethicalhour.com/energy-infrastructure/is-indias-transition-to-clean-energy-achievable-or-an-empty-promise.

Hall, S. 2023. "The Climate Disaster Strikes: What the Data Say." *Nature News*, December 13, 2023. Accessed October 14, 2024. https://www.nature.com/articles/d41586-023-03910-w.

Herrington, R. 2021. "Mining Our Green Future." *Nature Reviews Materials* 6, no. 6 (May 24): 456–8. https://doi.org/10.1038/s41578-021-00325-9.

Hilton, L. 2024. "China-US Engagement Critical to Global Climate Action." China.org.cn, January 28, 2024. Accessed October 14, 2024. https://e360.yale.edu/features/china-renewable-energy.

IEA. 2022. "Coal 2022—Analysis." Accessed July 11, 2024. https://www.iea.org/reports/coal-2022.

"India Eyes Africa's Vast Resources to Meet Rising Energy Demand." 2016. *Deccan Chronicle*, January 23, 2016. Accessed October 14, 2024. https://www.deccanchronicle.com/amp/economy/230116/india-eyes-africa-s-vast-resources-to-meet-rising-energy-demand.html.

"India Holds the Key to Hitting Global Climate Change Targets." 2023. *World Economic Forum*, January 19, 2023. Accessed October 14, 2024. https://www.weforum.org/agenda/2023/01/india-holds-the-key-to-hitting-global-climate-change-targets-here-s-why.

Intergovernmental Panel on Climate Change (IPCC). 2023. *Climate Change 2021—The Physical Science Basis*. June 29, 2023. https://doi.org/10.1017/9781009157896.

Kahn, M. E., and M. J. Kotchen. 2011. "Business Cycle Effects on Concern about Climate Change: The Chilling Effect of Recession." *Climate Change Economics* 2, no. 3 (August): 257–73. https://doi.org/10.1142/s2010007811000292.

Keane, P. 2020. "How the Oil Industry Made Us Doubt Climate Change." BBC News, September 19, 2020. Accessed October 14, 2024. https://www.bbc.com/news/stories-53640382.

Khan, A. 2012. "How Al Gore Galvanized the Climate Change Movement—on Both Sides." PBS, October 23, 2012. Accessed October 14, 2024. https://www.pbs.org/wgbh/frontline/article/how-al-gore-galvanized-the-climate-change-movement-on-both-sides.

Krajick, K. 2018. "James Hansen's Climate Warning, 30 Years Later." *State of the Planet*, June 26, 2018. Accessed October 14, 2024. https://news.climate.columbia.edu/2018/06/26/james-hansens-climate-warning-30-years-later.

Krange, O., B. P. Kaltenborn, and M. Hultman. 2021. "'Don't Confuse Me with Facts'—How Right Wing Populism Affects Trust in Agencies Advocating Anthropogenic Climate Change as a Reality." *Humanities and Social Sciences Communications* 8, no. 1 (October 29). https://doi.org/10.1057/s41599-021-00930-7.

Kulkarni, S. 2022. "Reversing Climate Change with Geoengineering." Science in the News, January 3, 2022. Accessed October 14, 2024. https://sitn.hms.harvard.edu/flash/2022/reversing-climate-change-with-geoengineering/2022.

Lindsey, R. and L. Dahlman. 2024. "Climate Change and Global Temperature." Climate.gov, January 18, 2024. Accessed October 14, 2024. https://www.climate.gov/news-features/understanding-climate/climate-change-global-temperature.

Malakoff, D. 2016. "Trump Picks Prominent Climate Skeptic as EPA Chief." Science.org, December 7, 2016. Accessed October 14, 2024. https://www.science.org/content/article/trump-picks-prominent-climate-skeptic-epa-chief.

Merriam Webster. 2011. "Mitigate." Accessed October 14, 2024. https://www.merriam-webster.com.

Moench, M. 2023. "Climate Change New: Anxiety Soars as Conditions Worsen." *Time*, November 22, 2023. Accessed October 14, 2024. https://time.com/6338759/climate-change-anxiety-google-search-trend.

Myers, C. 2016. "Changing Opinions on Climate Change." CNN, October 27, 2016. Accessed October 14, 2024. https://www.cnn.com/2016/08/24/opinions/chad-myers-climate-change-weather/index.html.

NASA. 2022. "Steamy Relationships: How Atmospheric Water Vapor Amplifies Earth's Greenhouse Effect." February 8, 2022. Accessed October 14, 2024. https://science.nasa.gov/earth/climate-change/steamy-relationships-how-atmospheric-water-vapor-amplifies-earths-greenhouse-effect.

"Neglected Crisis: Poverty, Terror and the Threat of Famine in the Lake Chad Basin." 2016. Theirworld, September 2016. Accessed October 14, 2024. https://theirworld.org/resources/neglected-crisis-poverty-terror-and-the-threat-of-famine-in-the-lake-chad-basin.

Otto, F. 2019. *Angry Weather: Heat Waves, Floods, Storms, and the New Science of Climate Change*. Vancouver: Greystone Books.

"President Bush Discusses Climate Change." 2008. White House Press Release, April 16, 2008. Accessed October 14, 2024. https://georgewbush-whitehouse.archives.gov/news/releases/2008/04/20080416-6.html.

Ritchie, H., and M. Roser 2020. "$CO_2$ Emissions". Our World in Data. Accessed October 14, 2024. https://ourworldindata.org/CO2-emissions.

Robinson, A., J. Lehmann, D. Barriopedro, S. Rahmstorf, and D. Coumou. 2021. "Increasing Heat and Rainfall Extremes Now Far Outside the Historical Climate."

*npj Climate and Atmospheric Science* 4, no. 1 (October 5). https://doi.org/10.1038/s41612-021-00202-w.

Rudiak-Gould, P. 2013. "'We Have Seen It with Our Own Eyes': Why We Disagree about Climate Change Visibility." *Weather, Climate, and Society* 5, no. 2 (April 1): 120–32. Accessed October 14, 2024. https://journals.ametsoc.org/view/journals/wcas/5/2/wcas-d-12-00034_1.xml.

Schwartz, J. 2018. "New York Sues Exxon Mobil, Saying It Deceived Shareholders on Climate Change." *The New York Times*, October 24, 2018. Accessed October 14, 2024. https://www.nytimes.com/2018/10/24/climate/exxon-lawsuit-climate-change.html.

Shaw, R., and S. Mukherjee. 2022. "The Development of Carbon Capture and Storage (CCS) in India: A Critical Review." *Carbon Capture Science & Technology* 2 (March): 100036. https://doi.org/10.1016/j.ccst.2022.100036.

Smith, W., and G. Wagner. 2018. "Stratospheric Aerosol Injection Tactics and Costs in the First 15 Years of Deployment." *Environmental Research Letters* 13: 124001. https://doi.org/10.1088/1748-9326/aae98d.

Supran, G., S. Rahmstorf, and N. Oreskes. 2023. "Assessing ExxonMobil's Global Warming Projections." *Science* 379, no. 6628.

Taylor, D. B. 2022. "One-Third of Americans Faced Extreme Weather in Recent Years, Survey Finds." *The New York Times*, April 7, 2022. Accessed October 14, 2024. https://www.nytimes.com/2022/04/07/climate/severe-weather-damages-us.html.

Thompson, C. 2019. "How 19th Century Scientists Predicted Global Warming." JSTOR Daily, December 17, 2019. Accessed October 14, 2024. https://daily.jstor.org/how-19th-century-scientists-predicted-global-warming.

Tschakert, P., N. R. Ellis, C. Anderson, A. Kelly, and J. Obeng. 2019. "One Thousand Ways to Experience Loss: A Systematic Analysis of Climate-Related Intangible Harm from around the World." *Global Environmental Change* 55 (March): 58–72. https://doi.org/10.1016/j.gloenvcha.2018.11.006.

USAID. 2023. "Focus on the Climate Crisis in Somalia: Somalia." *U.S. Agency for International Development*, September 12, 2023. Accessed October 14, 2024. https://www.usaid.gov/somalia/news/sep-08-2023-focus-climate-crisis-somalia.

Vaughan, C., S. Dessai, and C. Hewitt. 2018. "Surveying Climate Services: What Can We Learn from a Bird's-Eye View?" *Weather, Climate, and Society* 10, no. 2 (April 1): 373–95. https://journals.ametsoc.org/doi/10.1175/WCAS-D-17-0030.1.

Walsh, L., and T. Ormond-Skeaping. 2022. "The Cost of Delay: Why Finance to Address Loss and Damage Must Be Agreed at COP27." The Loss and Damage Collaboration, October 24, 2022. Accessed October 14, 2024. https://www.lossanddamagecollaboration.org/publication/cost-of-delay-why-finance-to-address-loss-and-damage-must-be-agreed-at-cop27.

Wood, K., and O. Ashford. 2023. "The Ocean Can Play a Bigger Role in Fighting Climate Change Than Previously Thought." World Economic Forum, September 29,

2023. Accessed October 14, 2024. https://www.weforum.org/agenda/2023/09/ocean-based-solutions-climate-crisis.

"Yale Experts Explain Climate Anxiety." 2023. Yale Sustainability, March 13, 2023. Accessed October 14, 2024. https://sustainability.yale.edu/explainers/yale-experts-explain-climate-anxiety.

Zhang, J. 2023. "Coal: The Ballast Stone of China's Energy Mix." National Energy Administration.

# 3

# Perspectives

Chapter 1 provided a background on the science that supports most of the research conducted on extreme weather and climate change. Chapter 2 offered an opportunity to look at problems, controversies, and proposed solutions to extreme weather events especially for vulnerable populations. Chapter 3 offers a unique chance to hear from various people in society including scientists researching the major topics, journalists communicating climate change and extreme events to the public and to policy makers, and people whose lives are affected on a daily basis by the changes in their environment. The authors provide an inside look at their perspectives from a first-person point of view. In particular, this chapter features a wildfire scientist, three journalists (Canadian, Chinese, and American), a climate change scientist who works with attribution studies looking at the link between global warming, emissions, and extreme weather events, and a citizen of Tanzania who has witnessed changes on Mt. Kilimanjaro's environment and changes to tourism and his society from extreme drought and flooding events.

## Canadian Wildfires and Climate Change: Canada Is Burning!

*Tricia Clarkson*

Usually when we think of catastrophic wildfires in Canada, we think of western Canada, specifically British Columbia, which is 235 million acres in size (double the size of California), where 66 percent of the province is forested. However, by

---

The aftermath of a forest fire in Jasper National Park, Alberta, Canada. Canada has experienced wildfires of increasing frequency and intensity in recent years, largely due to unusually warm and dry conditions. Smoke from Canada's wildfires has drifted to Europe and the United States, affecting air quality in a widespread area. (Brandon Smith/Dreamstime.com)

the end of 2023, wildfire blazes had engulfed parts of nearly all thirteen Canadian provinces and territories, drawing in federal and international firefighting resources to try to contain them (Reuters 2023). Quebec had an early start to the devastating 2023 wildfire season. Warm and dry conditions appeared in early May, setting the stage for June and July. On June 1, more than 120 fires were ignited by lightning in just one day. These conditions led to the largest single fire ever recorded in southern Quebec, which consumed 460,000 hectares (4,600 square kilometers) (Natural Resources Canada n.d.a.). This year was by far the most devastating wildfire season on record in Nova Scotia, with blazes burning through more than 25,000 hectares of land and 200 homes across the province on May 27th (CBC News 2023a).

The weather in Nova Scotia, which is almost entirely surrounded by the sea, is often characterized by high humidity and more moderate temperatures. But the Canadian Drought Monitor, a government agency, said the weeks leading up to the Nova Scotia fires were "significantly dry across southern parts of the Atlantic Region," with parts of the province seeing less than 50 percent of the average monthly precipitation (Kuebler 2023). The agency also noted that "a large pocket of severe drought" had been recorded for much of south and central Nova Scotia, which, combined with "significantly warm temperatures exceeding 30 °C (86 °F)," helped to create the ideal conditions for the wildfires (Kuebler 2023). "Climate change is increasing the frequency and intensity of wildland fires and creating longer fire seasons," Mike Norton, director general of the Canadian Forestry Centre, told reporters on June 5, 2023 (Kuebler 2023).

At the same time, firefighters were battling eastern blazes in Ontario, Quebec, and out-of-control fires in the western provinces of Alberta and British Columbia, forcing tens of thousands of people to evacuate in May. By June 12th, statistics from the Canadian Interagency Forest Fire Centre identified 2,467 active fires across the country. The agency expected 2023 to be Canada's worst wildfire season to date, and they were right. In July 2023 the Northwest Territories experienced their hottest temperatures ever recorded (Government of Canada n.d.).

What followed was an intense wildfire outbreak on August 16th that forced the 22,000 residents of Yellowknife to evacuate. At least four firefighters died while battling the blazes. "By the end of September, 2023, the NWT had recorded 303 fires, burning over four million hectares (40,000 square kilometers)," said NWT Fire Information Officer, Jessica Davey (Minogue 2023). In British Columbia, by December 7th, more than 2.84 million hectares (28,400 square kilometers) of forest and land burned, tens of thousands of people were forced to evacuate, and six firefighters died, making it the province's most destructive wildfire season on record (BC Wildfire Service n.d.). According to the International Institute for Sustainable Development, the climate crisis is outpacing British Columbia's ability to adapt (Terton and Qi 2023). The historic wildfire seasons in 2021, 2022, and 2023 are now officially the province's most expensive and most destructive wildfires on record—costing British Columbia over CAD17 billion, plus the cost of lives (CBC News 2023b). These fires were fueled by record-high temperatures and widespread drought conditions across the country (Natural Resources Canada n.d.a.).

Every year, Environment and Climate Change Canada (ECCC) ranks the top ten Canadian weather stories. "The fires dominated everything," says ECCC senior climatologist David Phillips. The blazes were happening from coast to coast to coast. By the end of wildfire season, over 184,493 square kilometers had burned, Phillips noted, making it the worst fire season in our country's history (Mortillaro 2023). Smoke from Quebec's wildland fires crossed the Atlantic Ocean, darkened skies in Europe and drifted over northern Portugal and Spain. The smoke's widespread reach was apparent in a plume of black carbon particles—commonly called soot—spanning eastward from North America and over more than 2,000 miles of the Atlantic Ocean, seen on NASA's satellite radar (Earth Observatory 2023). Canadian wildfires also had consequences for over 100 million Americans. Thick smoke hung in the air above many parts of the United States. New York City looked like a scene out of an apocalyptic movie. The *New York Post* blasted a front page headline: "Blame Canada!" (Mortillaro 2023). Even though the wildfires in Quebec were 500km away, my hometown of Peterborough, Ontario was also shrouded in smoke. City officials issued warnings for residents to stay indoors with windows closed. Mail delivery was put on hold and face masks were recommended to help protect residents from harmful toxins in the air.

Western Canada saw nineteen daily temperature records broken in July that fueled hundreds of out-of-control wildfires. It was the warmest summer in seventy-six years—dating back to the start of national record-keeping in 1948. The average temperature anomaly was +2.0 °C, beating the previous warmest in 2012 (Government of Canada n.d.). The year 2023 was the planet's warmest on record, according to an analysis by scientists from National Oceanic and Atmospheric Administration (NOAA)'s National Centers for

Environmental Information (NOAA 2024). Wildfire records shattered, with statistics for 2023 including:

- 184,493 square kilometers of woodland up in smoke across Canada.
- The amount of area burned was more than double the previous year.
- The area burned in Canada was more than the total area burned in the United States in the last five years.
- According to the Canadian Wildfire Evacuation Database, evacuations numbered 235,458 people (Government of Canada n.d.).

"The word 'unprecedented' doesn't do justice to the severity of the wildfires in Canada this year," says Yan Boulanger, research scientist in forest ecology at Natural Resources Canada. "From a scientific perspective, the doubling of the previous burned area record is shocking" (Natural Resources Canada n.d.a.). Research shows that changes in climate create warmer, drier conditions, leading to longer and more active fire seasons (NOAA n.d.).

Scientists from around the world conducted a study to determine the extent that climate change can be linked to these wildfires. Yan Boulanger and two other Canadian Forest Services scientists—Jonathan Boucher and Sandy Erni—were part of an international team of sixteen scientists involved in a rapid research study focused on Quebec's wildfire season. The team found that climate change more than doubled the likelihood of extreme fire weather conditions in Quebec (Natural Resources Canada n.d.a.). "Climate change is greatly increasing the flammability of the fuel available for wildfires because the trees, fallen trees and underbrush are all so dry," Yan explains. "This means that a single spark, regardless of its source, can rapidly turn into a blazing inferno." "There's no question, extreme weather, record-high temperatures and dry conditions caused by climate change intensified this year's wildfire crisis," said Boucher (Natural Resources Canada n.d.a.).

Over 150 fires that began in 2023 were still active in early spring of 2024. In previous years, only about fifteen continued to burn through the winter (Natural Resources Canada n.d.a.). By early May, huge wildfires had broken out in Alberta, British Columbia, and Manitoba with evacuation orders displacing thousands. Scientists at the World Resources Institute (WRI) and the University of Maryland calculated that the 2023 Canadian wildfires put 3.28 billion tonnes of heat-trapping carbon dioxide into the air, according to a study update published in *Global Change Biology*. "So when they (forests) burn, all that carbon that's stored within them gets released back into the atmosphere," said study lead author, James MacCarthy, a research associate with WRI's Global Forest Watch (Borenstein 2024).

The most heartbreaking wildfire in Canadian history occurred on July 22, 2024, when Canada's most precious jewel—Jasper National Park in Alberta—was ignited by wildfires. About 25,000 people fled Jasper National Park on July 22nd as a pair of wildfires drew dangerously close. Two days later, wind-whipped flames hundreds of feet high, overwhelmed firefighters and entered the community. The next day, on July 25th, the megafire entered the town of Jasper destroying approximately 50 percent of the town's structures and homes. A State of Emergency remains in place within the municipality and residents must be prepared to leave again at a moment's notice as a wildfire evacuation alert remains in effect (Snowdon 2024). As of August 13, 2024, 4,246 wildfires were burning throughout Canada (Canadian Wildland Fire Centre n.d.).

The International Panel of Climate Change (IPCC) concluded that if no specific actions were taken to reduce GHG emissions, global temperatures would be likely to rise between 1.4 and 5.8 °C from 1990 to 2100 (IPCC 2023). Area-burned projections, based on weather/fire-danger relationships, suggest a 75–120 percent increase in area burned by 2100, according to the Canadian and Hadley models (Flannigan et al. 2010). The 2023 wildland fire season was historic: 6,623 fires burned more than fifteen million hectares (57,915 square miles/150,000 square kilometers) of Canada's managed forests (Natural Resources Canada n.d.b.). Based on my calculation, a 75 percent increase in area burned means that at least 101,351 square miles (262,498 square kilometers) of area in Canada will burn to the ground by 2100 (57,915 square miles × 75 percent increase = 43,436 + 57,915 = 101,351 square miles/262,498 square kilometers).

What causes wildfires in Canada? A combination of lightning strikes and human activity (including fallen power lines, sparks from a train track, cigarette butts, barbeques, campfires and riding ATVs with exhaust systems that exceed 200 °C), combined with severe drought from climate change/global warming, cause wildfires. What can we do to stop wildfires from getting worse? Governments must stop the fossil fuel industry from pouring fuel on the flames. If Canada wants to be a strong climate leader, we need a strong federal cap on oil and gas emissions. Politicians have the power to slash carbon emissions to get us back on the path to a safer future, but they will only act if the public clearly demands it. An Oxford Economics advisory firm recently calculated that new green industries could be worth up to $10.3 trillion dollars to the global economy by 2050. We need to end fossil fuel subsidies and use the revenue gain to implement a green transition. In the meantime, we must choose all of the zero-emissions technologies available to us. This will improve our economy as well as our environment. In upcoming elections, we must vote for politicians that have climate change initiatives as their first priority, with the best plan for cleaner air. If not, we won't be around to breathe it.

## Further Reading

BC Wildfire Service. n.d. "Wildfire Season Summary." Accessed August 19, 2024. https://www2.gov.bc.ca/gov/content/safety/wildfire-status/about-bcws/wildfire-history/wildfire-season-summary.

Bhagat, Ali. 2024. "Wildfires Foreshadow the Looming Crisis of Internal Displacement." *The Globe and Mail*, August 12, 2024. Accessed October 14, 2024. https://www.theglobeandmail.com/opinion/article-wildfires-foreshadow-the-looming-crisis-of-internal-displacement.

Borenstein, S. 2024. "Canada's 2023 Wildfires Burned Huge Chunks of Forest, Spewing More Heat-Trapping Gas Than Planes." *Canada's National Observer*, June 27, 2024. Accessed October 14, 2024. https://www.nationalobserver.com/2024/06/27/news/canadas-2023-wildfires-burned-huge-chunks-forest-spewing-more-heat-trapping-gas.

Canadian Wildland Fire Centre. n.d. "National Wildland Fire Situation." Accessed August 13, 2024. https://cwfis.cfs.nrcan.gc.ca/report.

CBC News. 2023a. "Nova Scotia Saw Its Most Devastating Wildfire Season on Record in 2023." October 30, 2023. Accessed October 14, 2024. https://www.cbc.ca/news/canada/nova-scotia/nova-scotia-most-devastating-wildfire-season-ever-1.7010205.

CBC News. 2023b. "2023 Is Now Officially the Most Expensive, Most Destructive Wildfire Season on Record in B.C." September 27, 2023. Accessed October 14, 2024. https://www.cbc.ca/news/canada/british-columbia/wildfire-danger-reducing-1.6980041.

Earth Observatory. 2023. "Canadian Smoke Reaches Europe." NASA, June 26, 2023. Accessed October 14, 2024. https://earthobservatory.nasa.gov/images/151507/canadian-smoke-reaches-europe.

Flannigan, M., and M. Wotton, Canadian Forest Service, and Universities of Alberta and Toronto. 2010."Forest Fire Activity in Canada during the 2030s–2050s, Area Burned Projections: GCM Estimates." Wildland Fire Canada, October 6, 2010. Accessed October 14, 2024. https://sites.ualberta.ca/~wildfire/2010/PDFs/Mike%20Flannigan.pdf.

Government of Canada, Environment and Climate Change Canada (ECCC). n.d. "Canada's Top 10 Weather Stories of 2023." Accessed August 19, 2024. https://www.canada.ca/en/environment-climate-change/services/top-ten-weather-stories/2023.html#toc2.

IPCC. 2023. *Climate Change 2023: Synthesis Report. Contribution of Working Groups I, II, and III to the Sixth Assessment Report of the Intergovernmental Panel on Climate Change*. Geneva, Switzerland: IPCC. http://doi.org/10.59327/IPCC/AR6-9789291691647.

Kuebler, M. 2023. "How Canada's Wildfires Really Started." dw.com, June 12, 2023. Accessed October 14, 2024. https://www.dw.com/en/how-canadas-wildfires-really-started/a-65893778.

Minogue, S. 2023. "Yellowknife Begins Evacuation as Wildfires Approach." CBC News, August 16, 2023. Accessed October 14, 2024. https://www.cbc.ca/news/canada/north/nwt-wildfire-emergency-update-august-16-1.6938756.

Mortillaro, N. 2023. "Canada's Top 10 Weather Stories of 2023: Wildfires, Smoky Skies, a Record Hot Summer and More." CBC News, December 19, 2023. Accessed October 14, 2024. https://www.cbc.ca/news/science/2023-top-10-weather-stories-1.7063022.

National Oceanic and Atmospheric Administration (NOAA). 2024. "2023 Was the World's Warmest Year on Record, by Far." January 12, 2024. Accessed October 14, 2024. https://www.noaa.gov/news/2023-was-worlds-warmest-year-on-record-by-far.

National Oceanic and Atmospheric Administration (NOAA). n.d. "Wildfire Climate Connection." Accessed August 19, 2024. https://www.noaa.gov/noaa-wildfire/wildfire-climate-connection.

Natural Resources Canada. n.d.a. "Canada's Record-breaking Wildfires in 2023: A Fiery Wake-up Call." Accessed August 19, 2024. https://natural-resources.canada.ca/simply-science/canadas-record-breaking-wildfires-2023-fiery-wake-call/25303.

Natural Resources Canada. n.d.b. "Forest Fires." Accessed August 19, 2024. https://natural-resources.canada.ca/our-natural-resources/forests/wildland-fires-insects-disturbances/forest-fires/13143.

Reuters. 2023. "Canadian Wildfires Could Keep Burning through Winter, Minister Says." Reuters News, September 7, 2023. Accessed October 14, 2024. https://www.reuters.com/business/environment/canadian-wildfires-could-keep-burning-through-winter-minister-2023-09-07/.

Snowdon, W. 2024. "Anxiety, Relief as Jasper Residents Prepare for Return to Wildfire-ravaged Mountain Town." CBC News, August 13, 2024. Accessed October 14, 2024. https://www.cbc.ca/news/canada/edmonton/jasper-wildfire-alberta-1.7293176.

Terton, A., and J. Qi. 2023. "The Climate Crisis Is Outpacing B.C.'s Ability to Adapt." International Institute for Sustainable Development, October 26, 2023. Accessed October 14, 2024. https://www.iisd.org/articles/insight/climate-crisis-outpacing-british-columbia-adaptation.

*Tricia Clarkson is a climate change columnist, activist, and former co-chair for Peterborough Alliance for Climate Action who lives in Peterborough, Ontario, Canada. She writes weekly climate change columns for* The Peterborough Examiner *that are also published by several newspapers throughout Ontario, both in print and online. Her presentations at city council meetings have resulted in climate change initiatives that are reducing emissions throughout the city. Tricia, her husband and four sons have reduced their carbon footprint by incorporating zero-emissions, green technologies into their lifestyle. She and her family have all been affected by forest fires caused by climate change.*

## Communication Is a Climate Solution: Connecting the Dots Between Climate Change, Extreme Weather, and Priority Impacts

*Karen Florini*

Although nearly two-thirds of Americans are worried about climate change (Leiserowitz et al. 2023), it perennially ranks surprisingly low in surveys of national priorities (Tyson 2023). This low overall priority persists despite abundant scientific evidence linking climate change with higher-priority concerns such as health and the economy (Rocque et al. n.d.). In part, the disparity reflects the relative lack of media coverage that makes the connections along the full causal chain—one that goes from climate change to extreme weather (such as heat waves, drought, deluge, rising seas, and wildfires) and then to the resulting array of direct and indirect impacts on air quality, disease vectors, food and water supplies, coastal and inland flooding, and more. Although these impacts are occurring with increased severity and frequency (Climate Central 2024), most Americans say they hear about climate change in the media less than once a week (YPCC 2023). Most also say that they have not personally experienced climate impacts (YPCC 2023)—even though virtually everyone on the planet has recently been exposed to high-heat events with a clear fingerprint from climate change (Climate Central 2024).

To help media professionals and other climate communicators connect the dots, Climate Central creates science-backed tools, visuals, and analyses that highlight the linkages between climate change, weather, and impacts, routinely localizing the messages to drive them home. Climate Central also provides information on—though does not advocate for—a range of low-carbon solutions. The result is a steady stream of information confirming the accelerating impacts of climate change to a public that is increasingly aware but needs to be more fully engaged in addressing this challenge. Climate Central materials are publicly available online, in English and Spanish, and may be used without charge for any non-commercial or news-media purpose as long as credit is given. Media usage of the organization's resources has continually trended upward, exceeding 25,000 instances across print, digital, TV, and radio outlets in the United States during 2023—double the prior year's total. Climate Central's principal tools and programs include the following.

Each week, Climate Matters provides compelling graphics and analyses to more than 1,000 TV meteorologists in nearly 95% of US media markets (see Figure 3.1). These familiar faces incorporate the information into their daily weathercasts, showing millions of viewers at home how climate change is affecting their weather *locally* and *immediately*. Multiple peer-reviewed studies have shown that these communications increase viewer knowledge and concern about climate change (Myers et al. 2020; Feygina et al. 2020). The weekly bulletins are also distributed to 3,000 other media professionals and climate communicators in non-governmental organizations, governments, academia, and businesses, and simultaneously made available online in a searchable library (climatecentral.org/resources). Each bulletin focuses on a timely topic, ranging from extreme-weather impacts to low-carbon solutions, with brief bullets

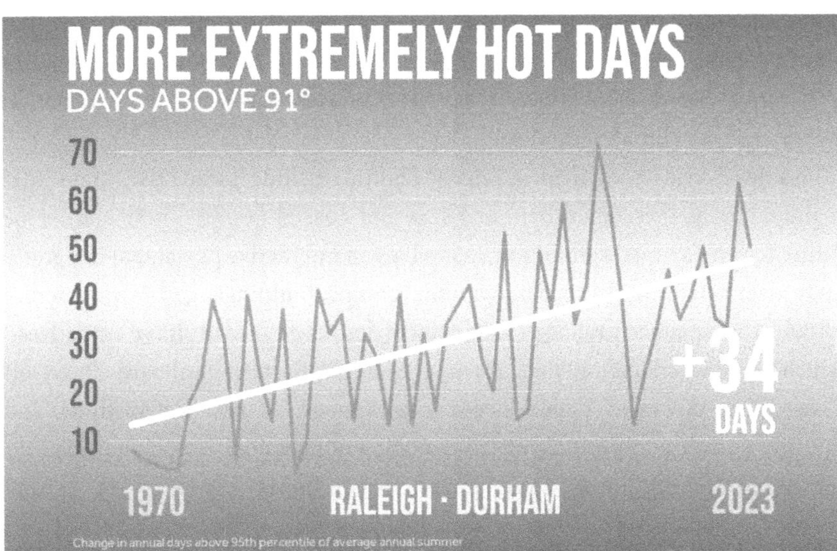


**Figure 3.1** Climate Central graphic showing the change in annual days above 95th percentile of average annual summer maximum temperatures (91.2°F), 1970–2023. (Data source: RCC-ACIS.org)

providing key contextual information. Climate Central also produces in-depth reports, with additional data and visuals, to provide background and more detailed scientific explanations, complementing weekly bulletins and helping users to develop their knowledge of these topics.

Using peer-reviewed methods, the Climate Shift Index quantifies the fingerprint of climate change everywhere, every day, via a single number on a 1-to-5 scale that indicates the extent to which the local temperature has been influenced by climate change. In the same way that temperatures and precipitation is routinely included in weather reports and forecasts, the Climate Shift Index provides a continuing reminder of the effects of climate change on daily temperatures (average, high, and low). From the global map at the Climate Shift Index online portal, users can navigate to specific information and downloadable screenshots for more than 10,000 locations worldwide (csi.climatecentral.org/climate-shift-index), as well as for sea-surface temperatures (which greatly affect tropical storms) (csi.climatecentral.org/ocean).

WeatherPower provides real-time information on the quantity of renewable electricity generated in every US state, county, media market, and congressional district. It combines information on installed wind and solar capacity with weather data and forecasts updated daily (see Figure 3.2). To put the resulting megawatt-hour numbers into context, the tool offers a variety of equivalencies, including number of homes powered, cellphones charged, and $CO_2$ tons avoided—showing how weather is powering a clean energy transition (weatherpower.climatecentral.org).

Sea level rise tools from Climate Central include interactive maps that highlight areas threatened by rising seas and coastal flooding now and in the future (coastal.climatecentral.org), as well as an interactive portal that integrates 100+ data sets on risks to demographic groups, buildings, and infrastructure (riskfinder.climatecentral.org). The newly launched *FloodVision* initiative—which is now collecting data along the entire Eastern Seaboard and Gulf Coast—uses advanced data collection techniques, AI-based algorithms, and intuitive software to produce photo-realistic visualizations of specific locations (climatecentral.org/floodvision). Such visualizations can provide crucial support for emergency managers who need to capture public attention and gain compliance with evacuation instructions when flooding may be imminent. They also help illustrate how rising seas increase flood vulnerability in coastal areas.

The Partnership Journalism program works with editors and reporters at local newspaper, TV, and radio outlets to co-produce original feature stories

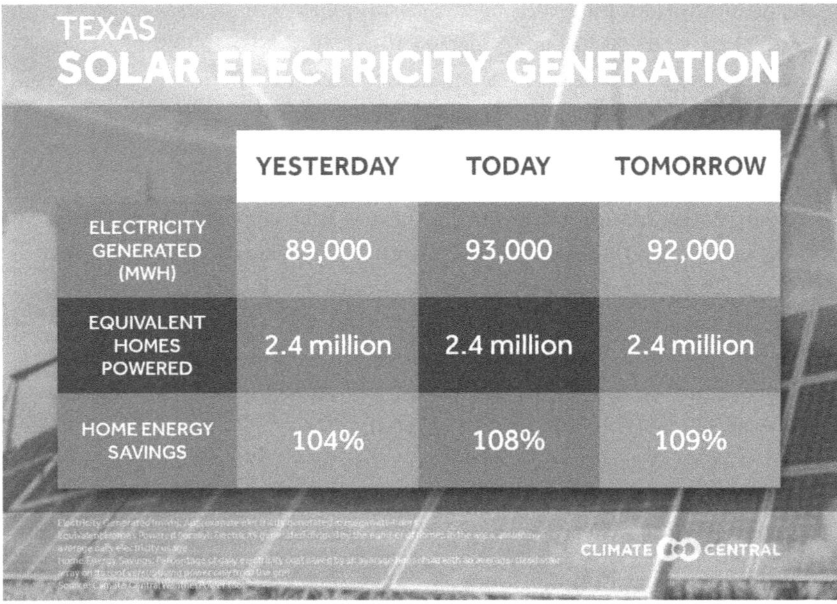


**Figure 3.2** Climate Central WeatherPower tool graphic showing data on solar electricity generation in Texas. Electricity Generated (mwh): approximate electricity generated in megawatt-hours; Equivalent Homes Powered (locally): electricity generated divided by the number of homes in the area, assuming average daily electricity usage; Home Energy Savings: percentage of daily electricity cost saved by an average household with an average-sized solar array on its roof versus using power only from the grid.

(climatecentral.org/partnership-journalism). Climate Central brings subject expertise, story ideas, scientific data, and graphics, while local partners provide local reporting, photography, and editing. All stories, which cover climate-related topics including impacts such as post-flood mold contamination (Gerstein and Brooker 2023) and solutions such as solar energy workforce development (Roedel and Miller 2023), are co-bylined and co-published.

## Further Reading

Climate Central. 2024. "2023: 5th-Hottest U.S. Year with Record Billion-Dollar Disasters." January 9, 2024. Accessed October 14, 2024. https://www.climatecentral.org/climate-matters/2023-5th-hottest-us-year-with-record-billion-dollar-disasters.

Feygina, I., T. Myers, B. Placky, S. Sublette, T. Souza, J. Toohey-Morales, and E. Maibache. 2020. "Localized Climate Reporting by TV Weathercasters Enhances Public Understanding of Climate Change as a Local Problem: Evidence from a

Randomized Controlled Experiment." *Bulletin of the American Meteorological Society* 101: E1092–E1100. https://doi.org/10.1175/BAMS-D-19-0079.1.

Gerstein, M., and J. Brooker. 2023. "Devastating Floods Leave Detroiters with Toxic Mold." *Partnership Journalism*, September 29, 2023. Accessed October 14, 2024. https://www.climatecentral.org/partnership-journalism/devastating-floods-leave-detroiters-with-toxic-mold.

Leiserowitz, A., E. Maibach, S. Rosenthal, J. Kotcher, E. Goddard, J. Carman, M. Ballew, M. Verner, J. Marlon, S. Lee, T. Myers, M. Goldberg, N. Badullovich, and K. Thier. 2023. "Climate Change in the American Mind: Beliefs & Attitudes." Yale University and George Mason University, Fall 2023. New Haven, CT: Yale Program on Climate Change Communication. Accessed October 14, 2024. https://climatecommunication.yale.edu/wp-content/uploads/2024/01/climate-change-american-mind-beliefs-attitudes-fall-2023.pdf.

Myers, T. A., E. W. Maibach, B. W. Placky, K. L. Henry, M. D. Slater, and K. L. Seitter. 2020. "Impact of the Climate Matters Program on Public Understanding of Climate Change." *Weather, Climate, and Society* 12: 863–76. https://doi.org/10.1175/WCAS-D-20-0026.1.

Rocque, R. J., C. Beaudoin, R. Ndjaboue, L. Cameron, L. Poirier-Bergeron, R.-A. Poulin-Rheault, C. Fallon, A. C. Tricco, and H. O. Witteman. n.d. "Health Effects of Climate Change: An Overview of Systematic Reviews." *BMJ Open* 11, no. 6: e046333. Accessed October 14, 2024. https://bmjopen.bmj.com/content/11/6/e046333.

Roedel, K., and E. Miller. 2023. "Nevada Shows States How to Build Workforce for Solar Energy Boom." Partnership Journalism, July 27, 2023. Accessed October 14, 2024. https://www.climatecentral.org/partnership-journalism/nevada-shows-states-how-to-build-workforce-for-solar-energy-boom.

Tyson, A. 2023. "What the Data Says about Americans' Views of Climate Change." Pew Research Center, August 9, 2023. Accessed October 14, 2024. https://www.pewresearch.org/short-reads/2023/08/09/what-the-data-says-about-americans-views-of-climate-change.

Yale Program on Climate Change (YPPC). 2023. "Yale Climate Opinion Maps 2023." *Yale Program on Climate Change Communication*, December 3, 2023. Accessed October 14, 2024. https://climatecommunication.yale.edu/visualizations-data/ycom-us.

*Karen Florini is Senior Advisor at Climate Central, where she engages with key strategic partners including NGOs, educators, governments, and businesses. Prior to joining Climate Central in 2017, Karen served as Deputy Special Envoy for Climate Change at the State Department. Previously she spent more than two decades at Environmental Defense Fund, working both on environmental health and on climate change. She earned a law degree at Harvard, where she was Editor-in-Chief of the* Harvard Environmental Law Review, *but now regards herself as a recovering lawyer.*

## Extreme Weather and Climate in the Late Little Ice Age

*Cary J. Mock*

Weather and climate extremes, such as atmospheric rivers, hurricanes, droughts, heat waves, cold air outbreaks, and snowstorms, have profound impacts on society, and at times are responsible for high amounts of monetary damage and sometimes causing fatalities. Atmospheric scientists study these extremes, based mostly from modern daily instrumental records that start around the mid-twentieth century (e.g., Marshall et al. 2024). Occasionally there are some records that extend back to the late nineteenth century in databases such as those archived by the NOAA of the United States. Although some extremes in recent decades may largely be related to anthropogenic-induced warming and climate change, the early-mid nineteenth century stands out as a unique period of natural climate variability also characterized with anomalous extreme events, and these aspects of natural variability must be clearly understood to attain a clear picture of future climate and extremes. The early-mid nineteenth century is known as the "Late Little Ice Age" (e.g., Wise and Dannenberg 2014; Slonosky 2018). The following examples describe some extremes and challenges that researchers face in studying the Late Little Ice Age, focusing on western North America and the North Pacific Ocean.

Long continuous reliable records of weather from daily data are scarce prior to about 1920, and they are even scarcer extending back into the 1800s for locations in places such as western North America. Figure 3.3 shows examples of a winter (December–February) rain day and rain amount time series for downtown San Francisco, California, updated from Mock (2012). The record started in late 1849, when the city grew quickly from the California gold rush. The rain day series shows a lot of variability with extremes occurring every two to eight years, but most extremes of very high and very low rain days occur prior to 1980. Prominent very low rain day winters include those of 1976 to 1977, 1876 to 1877, and 1863 to 1864. Both 1976 to 1977 and 1863 to 1864 are unusual dry winters related to a moderate El Niño, but perhaps related as well to abnormal displacement of broad-scale atmospheric circulation patterns. Very high rain day winters are evident for 1889 to 1890 and several winters in the early-mid 1900s, and the very strong El Niño winter of 1982 to 1983. Precipitation amounts for San Francisco generally agree with rain days, but some extreme winters differ in terms of the amounts. The dry winters of 1863 to 1864 and 1976 to 1977 are also evident but other winters also reveal similar dry amounts. Wet winters of 1861 to 1862, 1867 to 1868, 1877 to 1878, 1889 to 1890, and 1982 to 1983 are also evident. The winter of 1861 to 1862 is known for creating a vast inland sea and several large inland lakes in California. Although the downtown San Francisco record mostly exhibits reliable data quality, some older data before 1890 present some challenges, as at times changes in observation routines, observers, and instrument exposures occurred (Mock 2012). For example, Mock's experiences in comparing various San Francisco historical records with one another suggests the winter of 1849 to 1850 rivals the wet one of 1889 to 1890, and that the winter of 1850 to 1851 was likely not that abnormally dry.

A few Alaska examples reveal climate information extending back into the Late Little Ice Age. A long paleoclimate proxy from coralline algal skeletons, calibrated to actual sea-surface temperature values off Alaska, clearly reveal a colder Late Little Ice Age from about 1830 to 1880 (Williams et al. 2017)

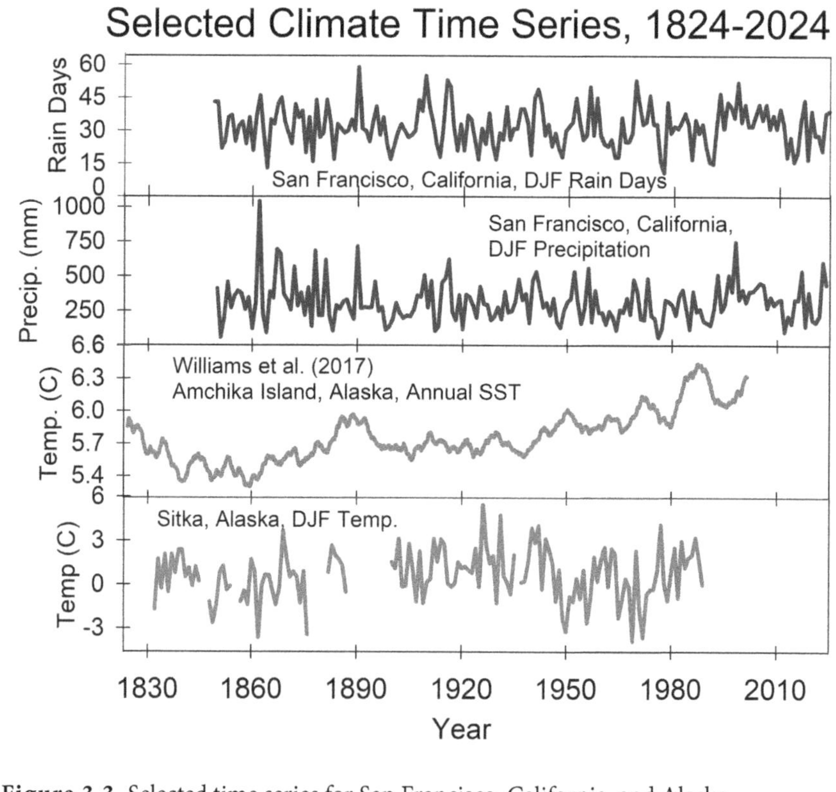


**Figure 3.3** Selected time series for San Francisco, California, and Alaska.

(Figure 3.3). Wise and Dannenberg (2014), using tree-ring proxies to study atmospheric circulation patterns in the North Pacific, reveal that this time frame was interesting concerning a lot of variability of the configuration of the jetstream—at times more west-east, and at times wavier with north-south aspects. Such variability may account for a higher probability of weather and climate extremes. A unique long temperature record from Sitka, Alaska, by the Russians (Parker 1981), extends back to the 1830s and also hints at generally colder temperatures, but the record is at times incomplete.

Understanding geographic variations of weather and climate in the Late Little Ice Age present even greater challenges, as this requires the combination of historical (non-instrumental), early instrumental, and paleoclimate proxy data such as tree-rings (Fritts, Lofgren, and Gordon 1981). At a minimum, annual dating is required, and sub-daily historical data are imperative. For locations such as Europe, studying geographic climate variations of the Little Ice Age has a long research tradition for many decades, while places such as western North America are still in very early research stages. Figure 3.4 shows an example of extreme weather highlights in mid-summer through fall 1849. Westward traveling 49ers passing through the Rocky Mountains and Intermountain West experienced abnormally cold and dry conditions, typical of the Late Little Ice Age, and they saw distant "perpetual snows" all summer on the peaks (Mock 1991). Other weather features include a very dry Pacific Northwest with large fires, as revealed by Hudson's Bay Company posts and ship logs. Although these fires could have been initiated by Native Americans, the dry weather undoubtedly was important for its spread and persistence. A somewhat stronger North American monsoon in the American Southwest, and eighty-two days of fog (July to October) at San Francisco were also evident.

Understanding atmospheric circulation weather patterns also presents a tough challenge for scholars. Tree-rings have been used to assess atmospheric circulation patterns in the North Pacific Ocean through statistical long-distance relationships of circulation to land networks of tree-ring sites (Wise and Dannenberg 2014). In mid-summer to fall in 1849, construction of detailed daily weather maps was possible from ship logs that include naval activity sailing in the California gold rush, and whalers sailing in the North Pacific through the Bering Straits into the Western Arctic. Generalized results reveal a dominance of the North Pacific subtropical high pressure system (Figure 3.4), similar to a pattern proposed by Fritts, Lofgren, and Gordon (1981). This subtropical high caused warm and dry conditions in the Pacific Northwest related to the fires, sinking air on its eastern edge over a cold ocean current caused conditions ideal

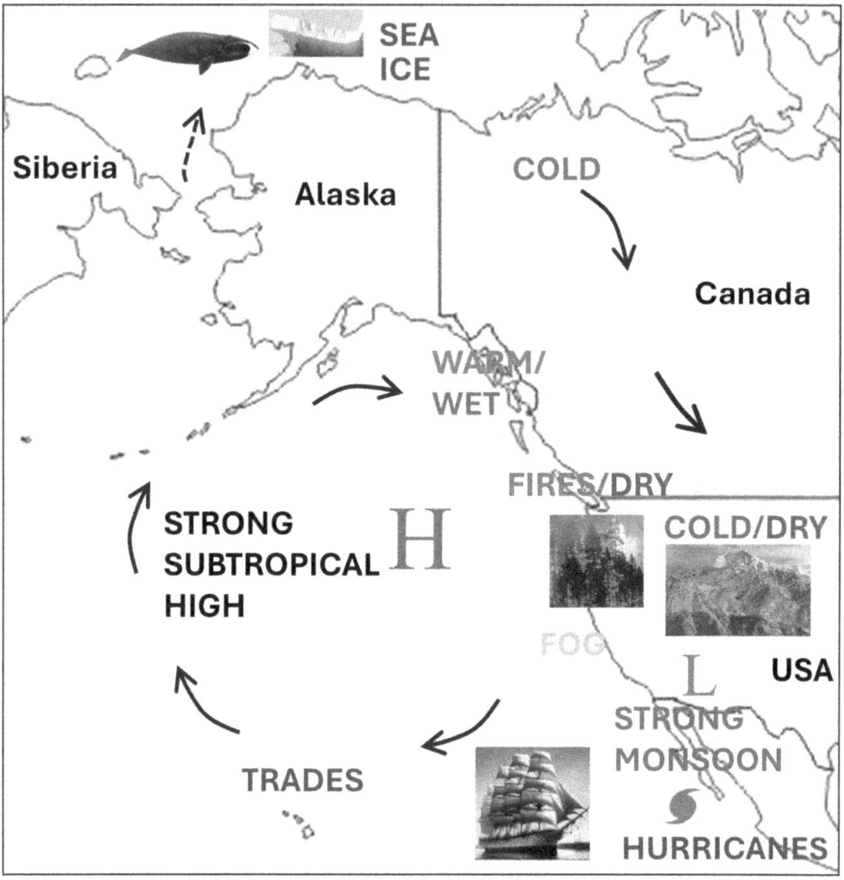


**Figure 3.4** 1849 Weather and impacts, mid-summer to mid-fall, for the North Pacific Ocean and western North America.

for California coastal fog formation. The high also brought somewhat warmer conditions from its western edge with southerly flow up through the Bering Straits, contributing to further northward sea ice extent than usual in mid-summer, very suitable for whalers. East of the subtropical high, northwesterly winds relate to the coldness in the Intermountain West into the Rockies, likely aided by cold air as well coming from western Canada. Ships traversing off west Mexico described up to ten possible tropical cyclones, which perhaps indicates a busy East Pacific hurricane season.

The examples described cover only a few selected locations for western North America and just one summer-fall season in the North Pacific. However, they reveal anomalous weather and climate extremes at times in the nineteenth century that exhibited more variability as compared to the modern record. Concerns of more weather and climate extremes will be critical in a world of future climate change and warming from anthropogenic sources. However, at times if unique circulation patterns that brought extremes in the Late Little Ice Age also occur in a future climate—it may result in some climate extreme surprises rarely witnessed to date. Challenges remain in historical climatology and paleoclimatology to study more of these past historical extremes, not just in western North America but also for other areas around the world as well. Citizen science projects involving the public are beginning to help out on the historical climatology and improving twentieth century reanalysis mapping products that extend back into the Late Little Ice Age (Slivinski et al. 2019).

## Further Reading

Fritts, H. C., G. R. Lofgren, and G. A. Gordon. 1981. "Reconstructing Seasonal to Century Time Scale Variations in Climate from Tree-ring Evidence." In *Climate and History*, edited by T. M. L. Wigley, M. J. Ingrain, and G. Farmer, 139-61. Cambridge, UK: Cambridge University Press.

Marshall, A. M., J. T. Abatzoglou, S. Rahimi, D. P. Lettenmaier, and A. Hall. 2024. "California's 2023 Snow Deluge: Contextualizing an Extreme Snow Year against Future Climate Change." *Proceedings of the National Academy of Science* 121, no. 20. https://doi.org/10.1073/pnas.2320600121.

Mock, C. J. 1991. "Historical Evidence of a Cold, Dry Summer during 1849 in the Northeastern Great Basin and Adjacent Rocky Mountains." *Climatic Change* 18, no. 1: 37-66. https://doi.org/10.1007/BF00142504.

Mock, C. J. 2012. "Instrumental and Documentary Evidence of Environmental Change." In *The Sage Handbook of Environmental Change*, edited by J. A. Matthews, P. J. Bartlein, K. R. Briffa, A. G. Dawson, A. de Vernal, T. Denham, S. C. Fritz, and F. Oldfield, 345-60. London: Sage Publications Ltd.

Parker, D. E. 1981. "Early Meteorological Observations for Sitka, Alaska." *Meteorological Magazine* 110: 161-3.

Slivinski, L. C., Gilbert P. Compo, Jeffrey S. Whitaker, P. D. Sardeshmukh, B. S. Giese, C. McColl, R. Allan, X. Yin, R. Vose, H. Titchner, J. Kennedy, L. J. Spencer, L. Ashcroft, S. Brönnimann, M. Brunet, D. Camuffo, R. Cornes, T. A. Cram, R. Crouthamel, F. Domínguez-Castro, J. E. Freeman, J. Gergis, E. Hawkins, P. D. Jones, S. Jourdain, A. Kaplan, H. Kubota, F. Le Blancq, T.-C. Lee, A. Lorrey, J. Luterbacher, M. Maugeri, C. J. Mock, G. W. K. Moore, R. Przybylak, C. Pudmenzky, C. Reason, V. C. Slonosky, C. A. Smith, B. Tinz, B. Trewin, M. A. Valente, X. L. Wang, C. Wilkinson, K. Wood,

P. Wyszyński. 2019. "Towards a More Reliable Historical Reanalysis: Improvements for Version 3 of the Twentieth Century Reanalysis System." *Quarterly Journal of the Royal Meteorological Society* 145: 2876–908. https://doi.org/10.1002/qj.3598.

Slonosky, V. C. 2018. *Climate in the Age of Empire*. Boston: American Meteorological Society.

Williams, B., J. Halfar, K. L. Delong, E. Smith, R. Steneck, P. A. Lebednik, D. E. Jacob, J. Fietzke, and G. W. K. Moore. 2017. "North Pacific Twentieth Century Decadal-scale Variability Is Unique for the Past 342 Years." *Geophysical Research Letters* 44: 3761–9. doi:10.1002/2017GL073138.

Wise, E. K., and M. P. Dannenberg. 2014. "Persistence of Pressure Patterns over North America and the North Pacific since AD 1500." *Nature Communications* 5, no. 4912: 3-4. http://doi.org/10.1038/ncomms5912.

*Cary J. Mock, PhD, is a professor at the University of South Carolina with research interests in historical climatology, synoptic climatology, and late-Quaternary paleoclimatology. His historical climatology interests focus mostly within the last few hundred years, including reconstructing extreme weather events and hazards and assessing historical impacts of climate on society. Geographically, his research covers most of North America but has also expanded internationally and includes East Asia, the Western Arctic, Belize, Bermuda, and Istanbul. Dr. Mock obtained his PhD. in geography at the University of Oregon in 1994. He was a visiting assistant professor at the University of Massachusetts in 1994-1995 and subsequently did a postdoc at the Quaternary Research Center, University of Washington. He joined the Department of Geography at the University of South Carolina in 1999. He recently coedited, with Scott Elias, the* Encyclopedia of Quaternary Science.

## Florida Coastal Resilience

*Carrie Shuman*

Right before Hurricane Ian was due to come through Florida, I tried to reassure one of my recently hired coworkers. Hailing from Ohio, she had never been through a hurricane and wasn't sure what to expect. Both of us were working at the Sanibel Captiva Conservation Foundation (SCCF) whose efforts focus on Sanibel and Captiva—two barrier islands off the Gulf Coast of Florida. I had been working as SCCF's Coastal Resilience Manager—focused on aspects of climate mitigation and adaptation including advocating for the use of nature-based solutions—for just over a year. Even prior to this, I knew to take hurricanes seriously and firmly believed in planning for them and not taking unnecessary risks. Even so, we were all shocked by just how much fury Ian unleashed, and my coworker was about to experience a potent Category 4, almost Category 5, hurricane as her very first hurricane.

Many of the people who had been living in the area for some years had 2004's Hurricane Charley for reference. Some likely felt that if they had weathered Charley successfully, they would do okay with Hurricane Ian. So, although Lee County called for a mandatory evacuation of hurricane surge evacuation Zones A and B on Tuesday morning, that may be why many people stayed. The county also issued the mandatory evacuation order less than 24 hours ahead of Ian's expected landfall in the state—a move that has since been the target of criticism and may have imposed additional challenges on those trying to leave the area.

However, even though it followed an eerily similar track to Charley, Ian was different in some incredibly dangerous ways. Similar to Charley, when Ian made landfall on Cayo Costa near Captiva the afternoon of September 28, 2022, it was just shy of a Category 5 storm. Ian underwent rapid intensification—where a tropical cyclone gains significant strength in a 24-hour period—just like Charley had almost 20 years prior. But, Ian dwarfed Charley with an eye approximately 35 nautical miles across compared to Charley's five, and with hurricane and tropical force winds spanning at least twice Charley's. And, while Charley barreled through Florida swiftly in about 7-8 hours, Ian meandered its way through, taking at least a day to fully cross the state.

This ultimately meant Ian had time to sit around and introduce record amounts of rain and storm surge, including up to 5 to 9 feet of storm surge above ground level (11-13 feet NAVD 88) across parts of Sanibel and Fort Myers Beach, contributing to the 149 deaths in Florida attributed to the storm, with most concentrated within Lee county and then radiating to other surrounding counties in southwest Florida (Borresen and Procell 2023). While everything was unfolding, local media was capturing the harrowing stories of people who had barely escaped drowning by crawling into attics and onto roofs or desperately clinging to trees for hours.

Post-storm, for most of us working at the non-profit, and like almost everyone else who lived and worked in the area, our normal lives came to a complete standstill. Our organization's temporary mission became helping people in the most immediate aftermath. For the first few weeks, we could only get to Sanibel and Captiva by boat, as the Sanibel Causeway—the only consistent way on and off the islands—had failed catastrophically in five places. The significant time we spent working on our facilities was punctuated with helping local residents muck out their houses, clear the jungle of downed vegetation in their yards, and pile the vast amounts of debris (including dry wall, furniture, and appliances) and their ruined personal belongings on the curb. I realized that amount of destruction we saw is almost impossible to convey to other people who haven't been through something comparable.

Ian—the costliest storm on record in Florida—was representative of the fingerprints climate change has been leaving on extreme weather events like hurricanes, including the increased occurrence of more intense storms (greater than Category 3) and the uptick in their likelihood to undergo rapid intensification, travel more slowly, and decay less rapidly over land (e.g., Klotzbach et al. 2022; Kossin 2018; Kossin et al. 2020; Li and Chakraborty 2020). Climate change is fast rewriting our new normal and storms like Ian no longer represent an occasional "fluke" event. It's also no wonder that Florida is also experiencing an insurance crisis as many companies are pulling out of the state or refusing to cover those properties at highest risk.

Prior to Ian, the concept of coastal resilience had been gaining traction across Florida as a whole. Legislation passed in 2021 formalized the appointment of a Chief Resiliency Officer for the state, the funding of the Resilient Florida Grant Program, and a mandate to the Florida Department of Environmental Protection to characterize the vulnerabilities of Florida communities to coastal flooding caused by sea level rise, storm surge, heavy rain events, and compound flooding that can occur when multiple sources of flooding converge at the same time. However, at the state scale, there is still only haphazard acknowledgment of the role climate change plays in the current and future increase in these and other impacts. And, while there has been a lot of momentum to adapt to flooding in Florida, little movement has been made to advance a broad state-wide effort to significantly reduce greenhouse gas emissions. This can be compared to treating the symptoms of a disease but spending little effort on the root cause.

During my own time working on Sanibel and Captiva, our island communities also saw positive momentum on multiple fronts in that same time period. Captiva finished multiple vulnerability assessments showing what parts of the island were most susceptible to flooding and when. Sanibel received a Resilient Florida grant to start pursuing their own. The City of Sanibel further articulated its starting commitment to exploring renewable energy and energy efficiency options, both of which reduce greenhouse gas emissions. My own work focused on partnering across our community to bolster four “pillars”: supporting both (1) adaptation planning and (2) greenhouse gas reduction (mitigation) efforts, (3) contributing to SCCF environmental policy efforts especially on the resilience front, and (4) engaging in significant education and outreach about our vulnerabilities, and balanced solutions including leveraging our natural systems.

Both islands also have had protections in place that improve their longer-term outcomes relative to other Florida communities. The progressive comprehensive “Sanibel Plan” was developed when the City of Sanibel was founded and prioritized the integrity of local ecosystems while also prizing “sanctuary-style” living that focused on harmony with nature. Over 70 percent of the island is held as conservation lands including by SCCF, the US Fish and Wildlife Service (as the J. N. “Ding” Darling National Wildlife Refuge), and the City of Sanibel. These lands contain marshes, mangroves, and beach dunes that have lots of benefits including reducing incoming wind and wave energy. Captiva, an unincorporated part of Lee County, also has its own Community Plan with many similar elements to Sanibel’s.

But post-storm, working on coastal resilience was initially an even more delicate pathway to walk in the weeks and months after the storm. Community members were reeling from both the experience and catastrophic damage and loss Ian imposed. As with most catastrophes, many people wanted to recapture what they had as quickly as possible, though it will likely take years for the region’s recovery. But eventually the storm amplified both the opportunities and the challenges around exploring ways to improve. For instance, based on the suggestion of a Sanibel councilman, I and a local resident helped form and have been co-chairing SanCap Resilience, a coalition of local citizens and sixteen organizations all dedicated to exploring solutions and providing resources for our post-Ian recovery and for ensuring our longer-term resilience against climate change impacts. Using the catalyzing energy after Ian to help establish this framework will hopefully lead to better coordination and unity in continuing to address the challenges of climate change.

At the same time, some challenges that were present prior to Ian have been amplified during the region’s recovery. For instance, development pressures have only continued to intensify. Captiva is currently fighting some recently adopted county ordinances that would favor one of the largest resorts on the island and undermine some of the height and density restrictions that were written into the Captiva Community Plan. This has implications for both islands, including because visitors to Captiva must first pass through Sanibel. There is strong consensus among Sanibel and Captiva residents that elevating buildings out of the reach of floodwaters is appropriate while making additional changes that could increase the number of people put in harms' way on a vulnerable barrier island, including potentially increasing evacuation time in the event of future storms, is not.

Captiva's efforts to convert from septic to sewer could also be jeopardized since the associated engineering studies are density- and capacity-based. Septic systems in many low-lying coastal areas in Florida, across the country, and more globally may stop working effectively as seas rise and allow lots of pollutants and excess nutrients to end up in local waters. Many of our local habitats and our wildlife face significant challenges like degraded water quality that will only make it harder for them to respond to the pressure of a changing climate.

Like many other places, we also struggle in Florida and in our region with conversations about retreat and moving away from the highest hazard areas. However, in the long run, mother nature may make some of those decisions for us, and perhaps in the more immediate future if we don't continue to look for meaningful and balanced ways to both mitigate and adapt. I've since started a new climate resilience position working on these issues a little more broadly in the region, and I remain optimistic that I can join my fellow colleagues in continuing to help southwest Florida do just that.

## Further Reading

Borresen, J., and C. Procell. 2023. "Ian's Deadly Path: A Visual Look at How the Hurricane Devastated Florida Region by Region." *USA Today*, January 10, 2023. Accessed November 13, 2024. https://www.usatoday.com/in-depth/graphics/2023/01/09/hurricane-ian-destruction-in-florida/10938513002.

Klotzbach, P. J., K. M. Wood, C. J. Schreck, S. G. Bowen, C. M. Patricola, and M. M. Bell. 2022. "Trends in Global Tropical Cyclone Activity: 1990–2021." *Geophysical Research Letters* 49, no. 6 (March 19, 2022). https://doi.org/10.1029/2021gl095774.

Kossin, J. P. 2018. "A Global Slowdown of Tropical-cyclone Translation Speed." *Nature* 558, 104–7. https://doi.org/10.1038/s41586-018-0158-3.

Kossin, J. P., K. R. Knapp, T. L. Olander, and C. S. Velden. 2020. "Global Increase in Major Tropical Cyclone Exceedance Probability over the Past Four Decades." *Proceedings of the National Academy of Sciences* 117, no. 22 (May 18): 11975–80. https://doi.org/10.1073/pnas.1920849117.

Li, L., and P. Chakraborty. 2020. "Slower Decay of Landfalling Hurricanes in a Warming World." *Nature* 587: 230–4. https://doi.org/10.1038/s41586-020-2867-7.

*Carrie Schuman holds a PhD from the University of Florida where she focused on Interdisciplinary Ecology, Fisheries and Aquatic Science, Tropical Conservation & Development, and a Masters Degree in Marine Science and Technology from the UMass School of Marine Sciences focusing on Zooplankton-produced Fluorescent Dissolved Organic Matter (FDOM). Her PhD research looked at oyster ecosystem services in the Guana Tolomata Matanzas National Estuarine Research Reserve—in particular, the oyster clearance rates in the reserve. She also incorporated how oystermen and fishermen use and think about oyster reefs in the St. Augustine region of Florida. She worked as a coastal resilience manager for the Sanibel Captiva Conservation Foundation.*

# A Climate Scientist's Perspective on Hurricanes, Climate Change, and Extreme Event Attribution

*Thomas Knutson*

## An Extreme Weather or Climate Event Just Happened ... Are Humans to Blame?

When an extreme weather or climate event occurs, I am sometimes asked by reporters whether this event was due to climate change (where they mean human-caused climate change, such as the global warming that is occurring due to the build-up of greenhouse gases in our atmosphere). The answer I give will depend on the particular event in question and can vary from saying that climate scientists are pretty sure the event was caused by human activity (i.e., burning fossils fuels over the past few centuries), to a much more nuanced answer along the lines of "we don't know at this point." Or somewhere in between. Let me try to explain why there can be such different answers depending on the event, with a few examples.

## Examples of Highly Confident Event Attribution

What leads to a more definitive answer about human activity causing an event? Let's look at an example. Take the recent record global mean temperature reported in 2023. It's the latest in a series of such warm records set in recent decades, with the last cold global mean temperature record set far back in the early 1900s. A few years ago, my colleagues and I wrote a paper on the attribution of one of these recent global warming records. We concluded that according to the latest climate models available, the event was essentially impossible without the warming influence of human-caused greenhouse gas increases.

The case is even stronger than that because it is supported by: (1) numerous past studies of climate change detection and attribution, (2) a clear trend over more than a century in the data (that we estimate is well outside the range of expected natural variability), (3) a recent warming trend appearing in the approximately 1,000-year paleoclimate reconstructions of northern hemisphere temperature from geologic records, (4) a high level of physical understanding of the link between greenhouse gas emissions and climate warming—as summarized in a series of IPCC assessment reports—and (5) the success of climate models at simulating the historical warming when forced by historical estimates of various climate forcing agents, including greenhouse gases, aerosols, and volcanic emissions.

Importantly, climate model projections made in 1989 have been compared with the actual evolution of Earth's climate since then, and the qualitative success of these simulations as predictions demonstrates that climate models projections can provide reliable information about the evolution of Earth's climate resulting from increasing greenhouse gases. At this point, the scientific debate about whether human-caused emissions of greenhouse gases have caused substantial global warming is basically over. And as a consequence, attribution statements about humans causing some climate events, like new record-high global mean temperature, have become very strong in some cases.

## Examples Where Human-caused Warming Likely Contributed to an Event

Let's now turn to some examples where the question whether humans caused an event requires a more nuanced statement. Take for example the cases of some recent highly destructive US landfalling hurricanes, like Katrina in 2005, Harvey in 2017, and Ian in 2022. Were these events caused by human-caused global warming? Rather than claiming that the entire hurricane event was caused by global warming (we will return to this question later), let's first consider the narrower question whether human-caused global warming contributed to these events in some way (i.e., made them more likely to occur, or contributed to the magnitude of the impacts).

Consider sea level rise. According to the IPCC, human-caused climate change has been the main driver of the increase in global mean sea level since at least 1971. Sea level has been rising historically where the mentioned hurricanes made landfall. It stands to reason that human-caused climate change has almost certainly made coastal inundation from hurricane-induced storm surges worse than they would have been, had an equivalent storm occurred in pre-industrial times prior to the sea level rise.

The IPCC has also concluded, with medium confidence, that tropical cyclone-related rainfall rates have increased due to human-caused climate change (tropical cyclone being a term that includes both tropical storms and hurricanes). IPCC's medium confidence resulted from supporting studies based on climate model simulations and on event attribution studies, while the IPCC noted there is still no clearly detectable trend in observed global tropical cyclone rainfall statistics. Such a change makes physical sense because a warmer atmosphere tends to hold more water vapor, so that a storm system like a hurricane, which feeds on moisture from its surrounding environment, gets an extra boost of moisture to convert into torrential downpours in and near the storm.

## Examples Where the Human Contribution to an Extreme Event Is More Uncertain

We can assert that coastal flooding from hurricane surges and rainfall rates out of hurricanes likely include some contribution from human-caused global warming. Also, model simulations and some observed trends suggest that Atlantic hurricanes are becoming more intense (about 3 percent stronger winds for a 2 °C global warming scenario) and that an increasing fraction of hurricanes have been undergoing rapid intensification as a result of global warming since 1980. Yet we haven't definitively attributed either observed Atlantic wind intensity or rapid intensification changes to greenhouse warming, mostly because of limitations of the data (only about four decades of reliable basinwide data exist for these metrics). This short data record means that a greenhouse warming-induced signal in these metrics since 1980 would be mixed together with—and possibly hidden by—strong Atlantic basin multidecadal variability. This multidecadal variability in turn is presumably not due to greenhouse gases but to some combination of natural variability and human-caused aerosol changes. This complicates attribution statements about Atlantic hurricane intensity and rapid intensification events—they become highly dependent on climate models, without the support of century-scale observed rising trends for example.

Finally, we need to be particularly careful with hurricane-related attribution statements because we still have not addressed the important question of how the *frequency* of hurricanes in various locations is being affected by human-caused global warming. The impacts of the hurricane changes mentioned above should scale with the total number of hurricanes occurring. Thus, to illustrate why frequency remains an important issue, consider that the time series of US landfalling hurricanes dating from 1900 or so do not show any significant trend—either in hurricanes overall or in very intense (Category 3 to 5) hurricanes. Also the approximately 1,000-year paleoclimate records of hurricane activity at several locations around the Atlantic basin do not show any clear rising trends during the recent century, in contrast to paleoclimate records of Northern Hemisphere temperature. This means that at a minimum it is very difficult to detect a human-caused imprint on the frequency of US landfalling hurricanes or major hurricanes.

Further, climate models tend to disagree among themselves on whether to expect more or fewer hurricanes or intense hurricanes in the Atlantic basin over the coming century. While models project an increase in the proportion of hurricanes that reach higher intensities over the coming century with global warming, many models also project decreasing overall frequency of hurricanes and of major hurricanes. As such, the future remains murky on changes in hurricane risk for the United States and surrounding regions, apart from the inexorable rise of sea level and the intensification of hurricane rainfall rates from the moister atmosphere. In addition, most of our Atlantic hurricane-related event attribution statements have to be strongly caveated at present because of these uncertainties in the science.

## Concluding Thoughts

Climate science has made much progress in documenting climate changes, simulating climate change with models, and increasing physical understanding of climate change (as embodied in climate models and the scientific theories which underpin them). But there are clearly many important remaining unanswered questions. Will there be more or fewer intense hurricanes in future decades? How much global warming will occur in future decades? How much sea level rise will occur over the next 50, 100, or even 500 years? Are there important lurking nonlinear changes or "tipping points" in the climate system that our current models and statistical methods are incapable of anticipating (i.e., the "flying blind" problem)? These and other challenges remain for the next generation of climate scientists to tackle.

Science is about seeking answers, challenging assumptions, and ultimately increasing our knowledge of the world and universe around us. If you are thinking of a career in climate science (or another area of scientific inquiry), I hope this essay will provide some flavor for the excitement of this field and the many important unanswered questions that remain.

## Further Reading

Knutson, T. R. 2019. "Understanding Extreme Weather Events in a Changing Climate." *Annual Review of Climate Change* 7, no. 1: 123–45.

*Thomas Knutson is a prominent climate scientist known for his research on extreme weather events and their connection to climate change. He has contributed significantly to the field, particularly in the study of hurricanes and their intensification. Knutson's work emphasizes the importance of understanding climate variability and its impact on human health and ecosystems. His research has been cited extensively in scientific literature, and he continues to be an influential voice in climate science. Tom Knutson is a Senior Scientist at NOAA's Geophysical Fluid Dynamics Laboratory. The views and opinions expressed here are his alone and do not represent official views of NOAA or the US government.*

# Human Decisions Altering Wildfire Courses

*Sarah Sayedi*

Wildfires are integral to many ecosystems, playing a crucial role in maintaining ecological balance and promoting biodiversity. However, in the face of human-induced climate change, wildfires have become a significant environmental challenge. As climate change transforms ecosystems, conditions such as drier weather and higher temperatures are becoming more common, turning more vegetation into potential fuel and thus increasing fire activity, especially in regions where vegetation growth is not limited by climate (Bowman et al. 2020).

These wildfires are intricately connected with climate change in a complex relationship that transcends simple cause and effect. Tipping elements in the Earth's system, when altered by climate change, can undergo dramatic shifts. Upon reaching critical thresholds, they can escalate global warming through mechanisms like releasing additional greenhouse gases or altering surface albedo. This results in a feedback loop where escalating global warming further intensifies wildfire dynamics (Wang et al. 2023). An example is the arctic and boreal regions, where rising temperatures are leading to an increase in fire activity. These fires not only contribute to thawing permafrost and releasing significant carbon stocks but also set the stage for an increased likelihood of future wildfires, illustrating the cyclical interplay between wildfires and climate change (Abbott et al. 2016; Gibson et al. 2018).

Other human factors have further complicated this relationship. Activities like land development, fire suppression tactics, and changes in natural vegetation have significantly altered wildfire landscapes. Many fire regimes are now heavily influenced by human actions, affecting different aspects of fire regimes such as frequency and extent of fires (Parisien et al. 2016). These altered conditions are conducive to invasive species, which disrupt global ecological integrity by outcompeting native vegetation, leading to biodiversity loss and changing the natural dynamics of fire in these ecosystems (Bishop et al. 2020). Additionally, increased human presence in areas like the Wildland-Urban Interface (WUI) has led to more ignition sources, with human-started fires being a major factor (Radeloff et al. 2005). Notable examples include Fourth of July firework displays in the United States which has led to the most human-started fires in the country (Balch et al. 2017).

The impact of wildfires extends far beyond the immediate environmental effects, significantly affecting societies globally. Physical and mental health, particularly among vulnerable groups like children and women, are profoundly impacted (Zhu et al. 2024). These effects are not confined to the immediate vicinity of the fires; for instance, toxic smoke from Canadian wildfires has been known to reach as far as New York City (Owens 2023). Economically, wildfires pose a significant threat to industries like timber, which is valued at approximately US$1.5 trillion annually. Over the past two decades, wildfires have devastated vast areas of timber-producing forests, with effects rippling across international borders (Bousfield, Lindenmayer, and Edwards 2023). Moreover, wildfires have indirect effects that manifest over time. The smoke from the 2020 Californian wildfires, for example, led to increased cloud formation across Europe, showcasing the global influence of these events (Mamouri et al. 2023). In the Himalayas, intensified wildfires release black carbon emissions, accelerating glacier melting and increasing the frequency of glacier-related disasters like avalanches and glacial lake outbursts (Li et al. 2016).

The recent emergence of unprecedented megafires in regions like Western America and Australia, characterized by their vast scale and intensity, underscores the compounded effects of climate change and the neglect of traditional land management practices of Indigenous communities (Bowman et al. 2020). Integrating these practices is essential in developing effective wildfire mitigation strategies. However, as climate change accelerates, managing wildfires has become increasingly complex. For example, in Nepal, community-led reforestation efforts aimed at restoring forests and mitigating natural hazards have paradoxically increased wildfire risks due to the dry conditions induced by climate change (Pokharel et al. 2023). This highlights the delicate balance needed in land management strategies.

The increasing intricacy of wildfires necessitates advanced and diverse mitigation strategies. It is vital to blend various viewpoints and expertise into the decision-making process and ensure the efficient conversion of scientific findings into practical policies. At regional and local levels, the insights of Indigenous communities are invaluable. Their deep understanding of local ecosystems contributes significantly to sustainable land and fire management practices (Fletcher et al. 2021). Public education is also a critical component of a comprehensive strategy, raising awareness about the environmental impacts of actions and fostering proactive responses to wildfire risks. This multifaceted approach, which combines scientific expertise, policy innovation, community involvement, and educational initiatives, forms a robust framework to address the complex challenges posed by wildfires in the era of climate change. It emphasizes the need for diverse voices and swift action in bridging the gap between science and policy (Sayedi et al. 2024).

Mitigating wildfire impacts globally requires a rapid and significant reduction in greenhouse gas emissions, crucial due to the numerous tipping points in Earth's systems sensitive to high emissions levels. This reduction is essential not only to lessen adverse effects but also to improve our ability to handle wildfire impacts, considering the significant influence of climate on fire regimes. Central to this strategy is a reexamination of forests as carbon reservoirs, particularly in light of evolving fire patterns, and the strategic management of forest fuels to balance risk reduction with ecological preservation. Intensifying efforts to curb greenhouse gas emissions enhances our chances of reducing the detrimental effects of wildfires, underscoring the need to adopt all feasible clean energy technologies (Abbott et al. 2023).

## Further Reading

Abbott, B., C. Abrahamian, N. Newbold, P. Smith, M. Merritt, S. S. Sayedi, J. Bekker, M. Greenhalgh, S. Gilbert, M. King, N. Zimmermann, and C. Breyeret. 2023. "Accelerating the Renewable Energy Revolution to Get Back to the Holocene." *Earth's Future* 11, no. 9. https://doi.org/10.1029/2023EF003639.

Abbott, B. W., J. B. Jones, E. A. G. Schuur, F. S. Chapin III, W. B. Bowden, M. S. Bret-Harte, H. E. Epstein, M. D. Flannigan, T. K. Harms, T. N. Hollingsworth, M. C. Mack, A. D. McGuire, S. M. Natali, A. V. Rocha, S. E. Tank, M. R. Turetsky, J. E. Vonk, K. P. Wickland, G. R. Aiken, H. D Alexander, R. M. W. Amon, B. W. Benscoter, Y. Bergeron, K. Bishop, O. Blarquez, B. Bond-Lamberty, A. L. Breen, I. Buffam, Y. Cai, C. Carcaillet, S. K. Carey, J. M. Chen, H. Y. H. Chen, T. R. Christensen, L. W. Cooper, J. H. C. Cornelissen, William J. de Groot, Thomas H. DeLuca, E. Dorrepaal, N. Fetcher, J. C. Finlay, B. C. Forbes, N. H. F. French, S. Gauthier, M. P. Girardin, S. J. Goetz, J. G. Goldammer, L. Gough, P. Grogan, L. Guo, P. E. Higuera, L. Hinzman, F. S. Hu, G. Hugelius, E. E. Jafarov, R. Jandt, J. F. Johnstone, J. Karlsson, E. S. Kasischke, G. Kattner, R. Kelly, F. Keuper, G. W. Kling, P. Kortelainen, J. Kouki, P. Kuhry, H. Laudon, I. Laurion, R. W. Macdonald, P. J. Mann, P. J. Martikainen, J. W. McClelland, U. Molau, S. F. Oberbauer, D. Olefeldt, D. Paré, M.-A. Parisien, S. Payette, C. Peng, O. S. Pokrovsky, E. B. Rastetter, P. A. Raymond, M. K. Raynolds, G. Rein, J. F. Reynolds, M. Robards, B. M. Rogers, C. Schäde, K. Schaefer, I. K. Schmidt, A. Shvidenko, J. Sky, R. G. M. Spencer, G. Starr, R. G. Strieg, R. Teisserenc, L. J. Tranvik, T. Virtanen, J. M. Welker, and S. Zimov. 2016. "Biomass Offsets Little or None of Permafrost Carbon Release from Soils, Streams, and Wildfire: An Expert Assessment." *Environmental Research Letters* 11, no. 3: 034014. https://doi.org/10.1088/1748-9326/11/3/034014.

Balch, J. K., B. A. Bradley, J. T. Abatzoglou, R. C. Nagy, E. J. Fusco, and A. L. Mahood. 2017. "Human-started Wildfires Expand the Fire Niche across the United States." *Proceedings of the National Academy of Sciences* 114, no. 11: 2946–51. https://doi.org/10.1073/pnas.1617394114.

Bishop, T. B. B., R. A. Gill, B. R. McMillan, and S. B. St. Clair. 2020. "Fire, Rodent Herbivory, and Plant Competition: Implications for Invasion and Altered Fire Regimes in the Mojave Desert." *Oecologia* 192, no. 1: 155–67. https://doi.org/10.1007/s00442-019-04562-2.

Bousfield, C. G., D. B. Lindenmayer, and D. P. Edwards. 2023. "Substantial and Increasing Global Losses of Timber-producing Forest Due to Wildfires." *Nature Geoscience* 16, no. 12: Article 12. https://doi.org/10.1038/s41561-023-01323-y.

Bowman, D. M. J. S., C. A. Kolden, J. T. Abatzoglou, F. H. Johnston, G. R. van der Werf, and M. Flannigan. 2020. "Vegetation Fires in the Anthropocene." *Nature Reviews Earth & Environment*, 1, 500–15. https://doi.org/10.1038/s43017-020-0085-3.

Fletcher, M.-S., A. Romano, S. Connor, M. Mariani, and S. Y. Maezumi. 2021. "Catastrophic Bushfires, Indigenous Fire Knowledge and Reframing Science in Southeast Australia." *Fire* 4, no. 3: Article 3. https://doi.org/10.3390/fire4030061.

Gibson, C. M., L. E. Chasmer, D. K. Thompson, W. L. Quinton, M. D. Flannigan, and D. Olefeldt. 2018. "Wildfire as a Major Driver of Recent Permafrost Thaw in Boreal Peatlands." *Nature Communications* 9, no. 1: Article 1. https://doi.org/10.1038/s41467-018-05457-1.

Li, C., C. Bosch, S. Kang, A. Andersson, P. Chen, Q. Zhang, Z. Cong, B. Chen, D. Qin, and Ö. Gustafsson. 2016. "Sources of Black Carbon to the Himalayan–Tibetan Plateau Glaciers." *Nature Communications* 7, no. 1: Article 1. https://doi.org/10.1038/ncomms12574.

Mamouri, R.-E., A. Ansmann, K. Ohneiser, D. A. Knopf, A. Nisantzi, J. Bühl, R. Engelmann, A. Skupin, P. Seifert, H. Baars, D. Ene, U. Wandinger, and D. Hadjimitsis. 2023. "Wildfire Smoke Triggers Cirrus Formation: Lidar Observations over the Eastern Mediterranean." *Atmospheric Chemistry and Physics* 23, no. 22: 14097–114. https://doi.org/10.5194/acp-23-14097-2023.

Owens, B. 2023. "Why Are the Canadian Wildfires So Bad This Year?" *Nature* 618, no. 7965: 439–40. https://doi.org/10.1038/d41586-023-01902-4.

Parisien, M.-A., C. Miller, S. A. Parks, E. R. DeLancey, F.-N. Robinne, and M. D. Flannigan. 2016. "The Spatially Varying Influence of Humans on Fire Probability in North America." *Environmental Research Letters* 11, no. 7: 075005. https://doi.org/10.1088/1748-9326/11/7/075005.

Pokharel, B., S. Sharma, J. Stuivenvolt-Allen, S.-Y. S. Wang, M. LaPlante, R. R. Gillies, S. Khanal, M. Wehner, A. Rhoades, K. Hamal, B. Hatchett, W.-Y. Liu, S. Mukherjee, and D. Aryal. 2023. "Amplified Drought Trends in Nepal Increase the Potential for Himalayan Wildfires." *Climatic Change* 176, no. 2: 17. https://doi.org/10.1007/s10584-023-03495-3.

Radeloff, V. C., R. B. Hammer, S. I. Stewart, J. S. Fried, S. S. Holcomb, and J. F. McKeefry. 2005. "The Wildland–Urban Interface in the United States." *Ecological Applications* 15, no. 3: 799–805.

Sayedi, S. S., B. W. Abbott, B. Vannière, B. Leys, D. Colombaroli, G. G. Romera, M. Słowiński, J. C. Aleman, O. Blarquez, A. Feurdean, K. Brown, T. Aakala, T. Alenius, K. Allen, M. Andric, Y. Bergeron, S. Biagioni, R. Bradshaw, L. Bremond, E. Brisset, J. Brooks, S. O. Brugger, T. Brussel, H. Cadd, E. Cagliero, C. Carcaillet, V. Carter, F. X. Catry, A. Champreux, E. Chaste, R. D. Chavardès, M. Chipman, M. Conedera, S. Connor, M. Constantine, C. C. Mustaphi, A. N. Dabengwa, W. Daniels, E. De Boer, E. Dietze, J. Estrany, P. Fernandes, W. Finsinger, S. G. A. Flantua, P. Fox-Hughes, D. M. Gaboriau, E. M. Gayo, M. P. Girardin, J. Glenn, R. Glückler, C. González-Arango, M. Groves, D. S. Hamilton, R. J. Hamilton, S. Hantson, K. A. Hapsari, M. Hardiman, D. Hawthorne, K. Hoffman, J. Inoue, A. T. Karp, P. Krebs, C. Kulkarni, N. Kuosmanen, T. Lacourse, M.-P. Ledru, M. Lestienne, C. Long,

J. A. López-Sáez, N. Loughlin, M. Niklasson, J. Madrigal, S. Y. Maezumi, K. Marcisz, M. Mariani, D. McWethy, G. Meyer, C. Molinari, E. Montoya, S. Mooney, C. Morales-Molino, J. Morris, P. Moss, I. Oliveras, J. M. Pereira, G. B. Pezzatti, N. Pickarski, R. Pini, E. Rehn, C. C. Remy, J. Revelles, D. Rius, V. Robin, Y. Ruan, N. Rudaya, J. Russell-Smith, H. Seppä, L. Shumilovskikh, W. T. Sommers, Ç. Tavşanoğlu, C. Umbanhowar, E. Urquiaga, D. Urrego, R. S. Vachula, T. Wallenius, C. You, and Anne-Laure Daniau. 2024. "Assessing Changes in Global Fire Regimes." *Fire Ecology* 20, no. 18. https://doi.org/10.1186/s42408-023-00237-9.

Wang, S., A. Foster, E. A. Lenz, J. D. Kessler, J. C. Stroeve, L. O. Anderson, M. Turetsky, R. Betts, S. Zou, W. Liu, W. R. Boos, and Z. Hausfather. 2023. "Mechanisms and Impacts of Earth System Tipping Elements." *Reviews of Geophysics* 61, no. 1: e2021RG000757. https://doi.org/10.1029/2021RG000757.

Zhu, Q., D. Zhang, W. Wang, R. R. D'Souza, H. Zhang, B. Yang, K. Steenland, N. Scovronick, S. Ebelt, H. H. Chang, and Y. Liu. 2024. "Wildfires Are Associated with Increased Emergency Department Visits for Anxiety Disorders in the Western United States." *Nature Mental Health*, 2, 379–87. https://doi.org/10.1038/s44220-024-00210-8.

*Sarah Sayedi is a Postdoctoral fellow at the Global Research Institute at the College of William and Mary. She holds a doctorate from Brigham Young University and a Masters degree from University of Tehran in environmental management. Her research focuses on assessing the risk of dangerous change in Earth systems (permafrost collapse, global wildfire, and water security). Her thesis was entitled "Modelling Hydrological Ecosystem Services for Different Land Use and Climate Change Scenarios." Sara's research interests also include global wildfire regime changes due to climate change, risk assessment, and environmental applications of Remote Sensing and Geographic Information Systems.*

## Climate Change in China

*Xinying Chang*

Climate change stands as one of the most pressing ecological and social challenges of our time. China has a typical monsoon climate with diverse climate types and significant regional differences. Similar to global climate change, China has experienced a notable warming trend, with the climate risk index showing an upward trend, making it a severe and urgent challenge. China is vulnerable to the impacts of climate change, as evident by its ranking of sixty-first out of 181 countries in the 2020 ND-GAIN Index. It is also home to sixteen of the twenty global regions most vulnerable to climate change. China's weather bureau described the country as "a sensitive region in global climate change." China's climate has been warming since the 1950s, with the average ground temperatures in China rising much more quickly than the global average. The China Climate Bulletin 2022 reported that the average temperature in 2022 was 10.51 °C, which was 0.62 °C higher than the 1991–2020 average. The average ground temperature reached 107.1 °C in 2023, which was the highest on record since 1951 ("China Climate" n.d.). The national average precipitation was 606.1 mm in 2023, 5 percent less than the 1991 to 2020 average, the lowest since 2012. The long-term trend seemed to be stable but intra-annual changes were significant as Spring and Fall precipitation rates declined 3.2 millimeters (mm)/year and 3.6 mm/year respectively while summer precipitation increased.

According to a report released by China's Ministry of Natural Resources, China's coastal sea levels have experienced an average annual increase of 4 mm between 1993 and 2022, exceeding the global average sea level rise of 3.4 mm. As a result, the coastal sea levels in 2022 were 94 mm higher than the long-term average observed from 1993 to 2011. This accelerated rise is attributed to a combination of factors, including rising water temperatures, the melting of glaciers and polar icecaps, and decreasing atmospheric pressure. With more than 40 percent of China's population and most of its industry located in low-elevation coastal areas, China is one of the most vulnerable regions in the world to sea level rise.

Extreme weather and climate events have become more frequent, affecting a wide area with increased intensity. The years from 2015 to 2022 were the eight warmest years on record. China Daily, a state-run media, reported that China's weather broke multiple records in 2023. Across the country 127 weather stations recorded their highest daily maximum temperatures. In early summer 2023, North China and the Huanghuai region experienced five consecutive high-temperature heat waves. It was estimated that heatwave-related mortality was as high as about 50,000 deaths in 2022. However, December 2023 marked Beijing's longest cold wave since records began in 1951. Although average rainfall in 2023 was below the long-term average, 55 national weather stations recorded their highest daily rainfall, according to the National Climate Center. The size of the population exposed to extreme rainfall events increased substantially in recent years. Gansu, Qinghai, and Sichuan had the largest increase in extreme rainfall events during 2000 to 2022, whereas Tibet, Xinjiang, and Sichuan recorded the largest increase in drought occurrence. Work by Zhou et al. (2017) suggests that seasonal droughts caused by climate change could lead to losses of approximately 8 percent in China's primary crops production.

Rising global temperatures due to climate change could also enhance the intensity of typhoons, making them more dangerous. In 2023, Typhoon Doksuri swept into southern China and then moved northward, triggering deadly heavy rains and floods in Beijing, Tianjin, and Hebei. Following closely, Typhoon Mawar also made its mark, breaking precipitation records in multiple cities in Guangdong, as well as Hong Kong and Macao.

China's development and climate change are deeply intertwined. During the "14th Five-Year Plan" period, China's ecological civilization construction has entered a critical period of taking carbon reduction as the key strategic direction, promoting the synergy of pollution and carbon reduction, promoting the comprehensive green transformation of economic and social development, and achieving a qualitative change in the improvement of ecological and environmental quality. President Xi Jinping has set ambitious targets to reach a peak in carbon emissions before 2030 and to strive for carbon neutrality by 2060. Significant strides have been made in renewable energy, with the costs of wind and solar power generation decreasing by 80 percent and 90 percent, respectively, over the past decade. Given that greenhouse gases disperse globally in the atmosphere, climate change is a truly global issue, requiring a coordinated global effort to effectively address it. China is taking active environmental protection measures, promoting clean energy and strengthening global cooperation to jointly address climate change and ensure a sustainable future for the Earth.

## Further Reading

Armstrong, M. 2021. "The Price of Solar Power Has Fallen by over 80% since 2010. Here's Why." World Economic Forum, November 4, 2021. Accessed October 14, 2024. https://www.weforum.org/agenda/2021/11/renewable-energy-cost-fallen.

"China Climate Bulletin 2023 Is Rolled Out." n.d. *China Meteorological News Press*. Accessed August 6, 2024. https://www.cma.gov.cn/en/news/NewsEvents/news/202402/t20240208_6068296.html.

China Daily Website—Connecting China Connecting the World. Accessed September 2, 2023. http://www.chinadaily.com.cn.

China Meteorological Administration. 2023. "China Climate Bulletin 2022." Accessed November 9, 2024. https://www.cma.gov.cn/en2014/news/News/202302/t20230224_5328330.html.

"China Warns That Its Temperatures Are Rising Faster Than Global Average." 2022. Reuters, August 4, 2022. Accessed October 14. https://www.reuters.com/world/china/china-warns-that-its-temperatures-are-rising-faster-than-global-average-2022-08-04.

Menghan, L. n.d. "China's Weather Breaks Multiple Records, Showcases Global Warming's Disastrous Effect." *Chinadaily.com.cn*. Accessed August 4, 2023. https://www.chinadaily.com.cn/a/202401/03/WS65950e98a3105f21a507a595.html.

Notre Dame Global Adaptation Initiative. n.d. "Country Index." University of Notre Dame. Accessed July 12, 2024. https://gain.nd.edu/our-work/country-index.

Shantu. 2023. "China's New Great Wall." *The Economist*, June 5, 2023. Accessed October 14. https://www.economist.com/china/2023/06/05/chinas-new-great-wall.

Stanway, D. 2023. "China Provinces Top List of World's Most Climate-vulnerable Regions - Data." Reuters, February 20, 2023. Accessed October 14. https://www.reuters.com/world/china/china-provinces-top-list-worlds-most-climate-vulnerable-regions-data-2023-02-20/.

Sun, Y, X. Zhang, Y. Ding, D. Chen, D. Qin, and P. Zhai. 2021. "Understanding Human Influence on Climate Change in China." *National Science Review* 9, no. 3 (June 29). https://doi.org/10.1093/nsr/nwab113.

World Bank Group. 2021. "Climate Risk Country Profile: China." Accessed October 14. https://climateknowledgeportal.worldbank.org/sites/default/files/2021-05/15395A-WB_China%20Country%20Profile-WEB_0.pdf.

Xu, X. 2023. "Beijing Records Longest Cold Wave in Modern History." CNN, December 25, 2023. Accessed October 14. https://www.cnn.com/2023/12/24/asia/beijing-cold-weather-china-intl/index.html.

Zhang, S, C. Zhang, W. Cai, Y. Bai, M. Callaghan, N. Chang, B. Chen, H. Chen, L. Cheng, H. Dai, X. Dai, W. Fan, X. Fang, T. Gao, Y. Geng, D. Guan, Y. Hu, J. Hua, C. Huang, H. Huang, J. Huang, X. Huang, J. S. Ji, Q. Jiang, X. Jiang, G. Kiesewetter, T. Li, L. Liang, B. Lin, H. Lin, H. Liu, Q. Liu, X. Liu, Z. Liu, Z. Liu, Y. Liu, B. Lu, C. Lu, Z. Luo, W. Ma, Z. Mi, C. Ren, M. Romanello, J. Shen, J. Su, Y. Sun, X. Sun, X. Tang, M. Walawender, C. Wang, Q. Wang, R. Wang, L. Warnecke, W. Wei, S. Wen, Y. Xie, H. Xiong, B. Xu, Y. Yan, X. Yang, F. Yao, L. Yu, J. Yuan, Y. Zeng, J. Zhang, L. Zhang, R. Zhang, S. Zhang, S. Zhang, M. Zhao, D. Zheng, H. Zhou, J. Zhou, Z. Zhou, Y. Luo, and Peng Gong. 2023. "The 2023 China Report of the Lancet Countdown on Health and Climate Change: Taking Stock for a Thriving Future." *The Lancet Public Health* 8, no. 12 (December 2023): e978–95. https://doi.org/10.1016/s2468-2667(23)00245-1.

Zhou, D., Y. Liu, Y. Feng, H. Zhang, Y. Fu, Y. Liu, and Q. Tang. 2022. "Absolute Sea Level Changes along the Coast of China from Tide Gauges, GNSS, and Satellite Altimetry." *Journal of Geophysical Research: Oceans* 127, no. 9 (September 2022): e2022JC018994. https://doi.org/10.1029/2022jc018994.

Zhou, S., W. Zhou, G. Lin, J. Chen, T. Jiang, and M. Li. 2017. "Adapting to Climate Change: Scenario Analysis of Grain Production in China." *China Agricultural Economic Review*, November 6, 2017. https://doi.org/10.1108/CAER-10-2016-0173.

*Xinying Chang is a researcher for China Central Agricultural Radio and Television School (CABTS). CABTS is an educational institution in China that serves farmers by providing education and training programs, agricultural technology extension, and scientific information. It operates through remote teaching or face-to-face modes under the auspices of the Ministry of Agriculture and Rural Affairs of the People's Republic of China. CABTS offers online courses through its online learning platform, benefiting millions of learners.*

## Extreme Weather in Arusha, Tanzania

*Thomas Meng'oru Laizer*

My name is Thomas Meng'oru Laizer, and I was born in 1979 in the Arusha region of Tanzania. I am part of the Maasai tribe and was born into a family of seven children. I was the third born after my two elder sisters. Most of my family are pastoralists and farmers as many Maasai depend on this type of work. We are the only tribe in the region, and in Tanzania as a whole, that are still very conservative in traditional culture. Maasai people also live in large parts of northern, central and southern Kenya. Our most traditional people live in houses that are either circular or bread-loaf-shaped and are made by women. The men make a circular fence we call an Enkang around the village to protect cattle at night from wild animals like the African lion. Some Maasai do leave traditional villages, receive some education, and even work in busy cities.

After primary school, I worked in mining for a short while. I was unable to continue with secondary school after 1996. In 2003 I started climbing Mt. Kilimanjaro as a porter as my family could not afford for me to go to secondary school or college. I ported with companies until 2013 taking tourists from all over the world on multi-day trips up to 19,710 ft. During the climb to the top you go through many changes of the mountain in ecology. I noticed a change in the plants and animal and the weather since I began climbing.

I understand that climate change is the change of weather conditions in a region from its regular long-term averages to abnormal situations. From my perspective, as I have been living in Arusha since I was born, the weather in Arusha has two regular rainy seasons, a short rainy season from November to January and a long rainy season from the beginning of March up to July.

Everything was very typical for many years. But in more recent years I have experienced some major changes which I think may be because of climate change! We have had extreme drought and extreme rain.

Here is what I have seen in my time in Arusha and climbing Mt. Kilimanjaro. Some days since 2010 have been hotter than I have ever experienced in my life. Due to higher heat many plants have disappeared. There have been many effects on many animals tame and wild during the drought season. Livestock animals have died for lack of pasture for grazing during times of heat and dryness. About five years ago we witnessed many pastoralists families lose their herd of cows from extreme drought. In the National Parks, a change in the rainy season has affected ecosystems mainly for the wildebeest migration from Serengeti to the Maasai Mara in Kenya. And this is due to the irregular rainy season that has affected that migration for some years now. Pastoralists suffered a lot as their herds of cows became reduced in size due to drought. But they also suffered during years like in 2023 when during the heavy rain season they lost their cows from floods.

Many local people lost their jobs in farming and tourism due to more unpredictable situations in the region. Without agriculture there is no food for those who are living in town. Many were let go from work positions or their salaries were lowered. Many people in general are suffering due to climate change as it has changed everything. Arusha has about 2.6 million people that in some way rely on farming, ranching, or tourism. It is known mostly though as a tourism city.

During my ten years on Mt. Kilimanjaro as a porter, I first saw the top of the mountain filled with unmelting ice from Lava Tower at 4600m in western Breach to the summit at 5895m. Now from the crater to the summit there is a rapidly melting ice cap. The M glacier used to cover the crater and the summit. The M glacier is about to disappear. When I first started porting it was like a skyscraper! In ten years, the melting of the ice cap on Kilimanjaro has been so rapid. Many tourists come to see the glacier. Sometimes tourists get stuck in the parks due to extreme rainfalls during a time that should be dry season. Climate change has made planning more difficult as the weather is more unpredictable.

In 2023, the abnormal rainfall started in September and it lasted all the way to April of 2024. It was daily and no people went to work. Many lost their lives including so many children who were trying to get to school to fight for their education but ended up losing their lives in floodwaters. People had no time to work during that time last year of heavy rain.

In an effort to help with climate change, the government of the United Republic of Tanzania through our late president John Mafuli, stopped any manufacturing of plastics. It was illegal if you were found with it. Also, every Saturday was to be the day to make sweeping (clean-up) everywhere in every region and it was compulsory. The government planned to plant trees everywhere in open areas to try to bring humidity to the soil. Our people hope that extreme weather will become less common for our children and their children as governments help to reduce the problem.

## Further Reading

Karimi, M. 2022. "Climate: WFP Helps Pastoralists in Kenya Adapt to Changing Weather: World Food Programme." UN World Food Programme, March 2, 2022. Accessed October 14, 2024. https://www.wfp.org/stories/climate-wfp-helps-pastoralists-kenya-adapt-changing-weather.

"Maasai Tribe." n.d. Maasai Tribe Facts, Language, Religion, Culture, Diet & Clothing. Accessed July 15, 2024. https://www.masaimara.travel/maasai-tribe-facts.php.

Shirima, K. C., and C. G. Mung'ong'o. 2020. "Agroecosystems' Resilience and Social-Ecological Vulnerability Index to Climate Change in Kilimanjaro, Tanzania." *Climate Change Impacts and Sustainability: Ecosystems of Tanzania*, 34–43. Egham, UK: CABI. https://doi.org/10.1079/9781789242966.0034.

Stadelmann, C., J. Fürst, T. Seehaus, T. Mölg, and M. Braun. 2020. "The State of Kersten Glacier and the Northern Icefield on Mt. Kilimanjaro." *EGU General Assembly*, March 23, 2020. https://doi.org/10.5194/egusphere-egu2020-9339.

Swilla, L., Z. Katambara, and M. Lingwanda. 2024. "Evaluating the Performance of Low-Impact Development Practices in Mitigating Runoff Floods in the Kinyerezi River Catchments in Dar Es Salaam, Tanzania." *Water Conservation Science and Engineering* 9, no. 2 (June 18). https://doi.org/10.1007/s41101-024-00271-x.

"Tanzania Records Extreme Weather Events in 2021 Caused by Climate Change." n.d. Xinhua. Accessed July 15, 2024. https://english.news.cn/africa/20220304/d1741e9901aa4c30b9d99adf145b3805/c.html.

*Thomas Meng'oru Laizer is a member of the Maasai tribe of Tanzania and Kenya. He worked as a porter for Climb Kili for ten years and has taken many international tourists to the 19,000+ feet summit of Mt. Kilimanjaro. He lives with his wife and two sons near the city of Arusha, Tanzania.*

# 4

# Profiles

The astronomy, chemistry, and physics of our planet that influence the Earth's climate have been a subject of thought since humans first noticed a change in seasons. Early cultures worldwide, such as the builders of Stonehenge in England, left behind evidence that they acknowledged cyclical climate shifts. Archeologists theorize that the massive stones line up with the sunrise on the summer solstice and sunset on the winter solstice. Such massive efforts were perhaps made to capture what the early people knew was a yearly occurrence. The placement of the monoliths highlights agricultural seasons.

Thousands of years later, brilliant astronomers like Copernicus, Galileo, and Milankovitch would understand the influence that the revolution and the tilt of the Earth had on the climate. Scientists such as those of the mid-1800s to the early-1900s made the connection that the atmosphere was composed of gases and that the Earth's energy budget, or the balance of incoming vs. outgoing solar radiation, kept the temperatures of the Earth conducive to life on Earth. Eunice Foote (1818–98) who experimented with carbon dioxide ($CO_2$) in glass jars, was the first scientist to conclude that certain gases warmed when exposed to sunlight. She warned that increased levels of $CO_2$ in the atmosphere could affect the Earth's climate. Her studies were the foundation of what is now understood as the greenhouse effect. Svante Arrhenius, winner of the Nobel Prize in Chemistry (1903), was the first to use the laws of chemistry and physics to estimate how much the addition of atmospheric $CO_2$ would increase Earth's surface temperature. The scientific work of these early climate scientists would

---

UN Secretary-General António Guterres has called on developed nations to pool resources, scientific knowledge, and technologies to make carbon neutrality a reality and has urged wealthier nations to further climate justice by increasing aid to those nations most vulnerable to the effects of climate change. (Richard Juilliart/ Dreamstime.com)

lead to the founding principles used by modern climate researchers to develop critical instruments and be part of extensive collaborative efforts to communicate climate change and the resulting extreme effects it would have on our weather.

## Individuals

### Richard Alley (1957–)

Richard Blane Alley was born on August 18, 1957. He received his PhD from the University of Wisconsin-Madison and is an Evan Pugh professor of Geoscience at The Pennsylvania State University. Dr. Alley is a geoscientist who uses ice cores from the Earth's cryosphere (areas covered in ice) to reconstruct past climates. He is particularly interested in finding evidence of when the Earth's system reached "tipping points" (environmental conditions resulting in climate change). He has been on the climate expert panel of the United Nations, World Meteorological Organization since 1992.

His writings include 240 peer-reviewed scientific publications concerning Earth's cryosphere and global climate change. His 1988 *Nature* article "*Timing of Abrupt Climate Change at the End of the Younger Dryas Interval from Thermally Fractionated Gases in Polar Ice*," based on climate proxy data from Greenland ice cores, concluded that the last Ice Age ended abruptly and violently. The publication prompted Vice President Al Gore to invite him to testify about abrupt climate change before Senate and Congress. Dr. Alley was an author of the second, third, and fourth Intergovernmental Panel on Climate Change (IPCC) assessment reports. He was a lead author of "Chapter 4: *Observations: Changes in Snow, Ice, and Frozen Ground*" for the Fourth Assessment Report which shared the 2007 Nobel Prize with Al Gore. In 2000, he published *The Two-Mile Time Machine: Ice Cores, Abrupt Climate Change, and Our Future*. Numerous climate change-related television documentaries featured him as an expert. Dr. Alley has given many public presentations and media interviews about the subject. Dr. Alley gave the Bjerknes lecture at the 2009 American Geophysical Union meeting titled "The Biggest Control Knob—Carbon Dioxide in Earth's Climate History." Dr. Alley has received a number of distinguished honors including the BVAA Foundation Frontiers of Knowledge Award and the Heinz Award. He is also a Fellow of the American Academy of Arts and Sciences, was named to the National Academy of Sciences, and is a Foreign Member of the Royal Society.

While it is true that he has published many break-through scientific discoveries about climate in peer-reviewed academic journals, he is also committed to the communication of science to a general audience. He has been given many compliments for his books, his talks, and his recorded lectures.

## Harold E. Brooks (1959–)

Harold Brooks is a meteorologist specializing in the relationship between tornado events and climate change. He is a summa cum laude graduate of William Jewell College, where he studied physics and mathematics in 1982. He earned his Master of Arts and Philosophy at Columbia University. In 1990, he earned his PhD in atmospheric sciences from the University of Illinois at Urbana-Champaign. In 1991, Brooks joined the National Severe Storms Laboratory (NSSL) as a research meteorologist. He was elected as a Fellow of the American Meteorological Society (AMS) in 2010 and a Fellow of the Royal Meteorological Society (RMetS) in 1996. Brooks received the Department of Commerce Silver Medal in 2002, three National Oceanic and Atmospheric Administration (NOAA) Research Outstanding Paper Awards, and the NOAA Administrator's Award in 2007.

Dr. Brooks contributed to the Physical Science Basis portion of the Fourth Assessment Report of the Intergovernmental Panel on Climate Change (IPCC). He is a Senior Research Scientist at NOAA's National Severe Storms Laboratory, where he has worked twenty plus years on the climatology of severe thunderstorms and tornadoes and the overlap between social science and weather.

His current research focuses on the relationship between global warming and tornadoes. He has published articles in high-impact journals concerning the possible link between a warming world and the frequency or likelihood of tornado-producing weather. He has found that high-resolution climate simulations of a warmer planet project an atmosphere meteorologically supportive of tornadoes, especially in the Northern Hemisphere's spring, primarily based on an increase in the frequency of high wind shear values in the lowest 1 kilometer. His research has been instrumental in tornado forecasting and extreme weather due to climate change. He collaborated on a 2014 paper in the journal *Science* that concluded that the annual frequency of tornadoes has not increased since 1954. However, the number of tornadoes on a given day has increased.

He and his research team have found that tornado variability may or may not relate to increased warming. They are focused on the greater concentration of tornadoes on fewer days. In fact, around 20 percent of tornadoes in the most recent decade have occurred on the three days with the most tornadoes that year. This is a 10 percent increase from the prior decade. Brooks also studies the effects of the US warning systems and forecast models for extreme weather and tornadoes. His research is instrumental in pinpointing the possibility of dangerous weather events. His projections for possible changes in tornado characteristics and frequency are important to policymakers, emergency managers, and the community. He aims to improve the relay of warning systems from forecasts, observations, computer models, and risk communicators to the public. He hopes the improvements will lessen the devastating impacts of extreme snowfall, violent tornadoes, hurricanes, widespread flooding, and catastrophic drought. He is a member of NOAA's Weather-Ready Nation, an organization that prepares communities for extreme weather, water, and climate events.

## Kerry Emanuel (1955–)

Kerry Andrew Emanuel was born on April 21, 1955. He received his PhD from the Massachusetts Institute of Technology (MIT), where he has been a faculty member since 1981. He is a professor of meteorology specializing in atmospheric convection and the physical laws of the Earth's system that intensify hurricanes. Dr. Emanuel is the Director of the Center for Meteorology and Physical Oceanography and the Program in Atmosphere, Oceans, and Climate. He is one of the world's experts in tropical meteorology and cumulus convection, the role of clouds, water vapor, and upper ocean mixing in climate regulation, and advanced methods of sampling the atmosphere to model atmospheric scenarios depending on anthropogenic greenhouse gas emissions. In 2006, he was included in *Time Magazine*'s "100 People Who Shape Our World."

In March 2008, he published a landmark paper titled "The Hurricane-Climate Connection" in the *Bulletin of the American Meteorological Society*, with the conclusion that higher global average temperatures are likely to increase the power but decrease the prevalence of hurricane and cyclone activity. In 2011, he and Daniel H. Rothman, a professor of geophysics at MIT, co-founded the MIT Lorenz Center. The center aims to attract the brightest minds in physics, applied mathematics, chemistry, biology, and Earth science.

It provides students and faculty with resources and opportunities to creatively investigate climate change based on the idea that climate science is incredibly dynamic. The Lorenz Center hosts an annual public outreach lecture, the John Carlson Lectures Series, to reveal the latest climate science discoveries to the general public. In 2012, Dr. Emanuel served as keynote speaker at a conference for Republican voters concerned about climate change. He was harassed with menacing emails, including threats of violence after the blog *Climate Depot* posted his email address.

He advocates incorporating nuclear energy into the US supply grid. In 2013, he coauthored an open letter to policymakers supporting nuclear power to avoid dangerous climate change. Nuclear power would add to the clean energy options. It can also provide the massive amounts of power needed for many highly populated energy intensive cities.

## António Guterres (1949–)

António Guterres, born in 1949, is a Portuguese politician and diplomat. Since 2017, he has served as the ninth Secretary-General of the United Nations. He was prime minister of Portugal from 1995 to 2002. In March of 2023 he called on developed nations to augment their efforts to reach net-zero. He agreed with the findings by climate scientists that humans are responsible for essentially all global heating over the preindustrial average during the last 200 years, that the rate of temperature rise in the last half-century is the highest in 2,000 years, and that the concentration of $CO_2$ is at its highest in at least two million years. He almost optimistically emphasized that the IPCC's AR6 report was a how-to-guide to defuse the climate time bomb. He called it "a survival guide for humanity." He stressed that the 1.5 °C limit set by the Paris Agreement was still achievable, but not without a "fast-track climate effort by every country and every sector and on every timeframe" (Guterres 2023).

Guterres introduced the Climate Solidarity Pact, a plan by which the highest emitting nations make extra efforts to cut emissions as quickly as possible. The pact also encourages wealthier countries to provide financial and technical support to emerging economies. Guterres clarified that developed nations needed to reach net-zero by 2040 and emerging economies by 2050. Nations could do that based on common but varied responsibilities depending on a country's circumstances. He called for the G20 (an intergovernmental forum composed of nineteen sovereign nations, the European Union, and the African

Union who work on international financial stability, climate change mitigation, and sustainable development) to join the effort to pool resources, scientific knowledge, and technologies to make carbon neutrality a reality.

The Secretary-General clarified that developed nations must not develop any new coal, and any existing coal must be phased out by 2030 and in developing nations by 2040. Net-zero electricity generation should be achieved by 2035 for all developed countries and 2040 for any remaining nations. The international community should cease all licensing or funding of new oil and gas, stop any addition or expansion of existing oil and gas reserves, transfer subsidies from fossil fuels to a fair energy modification, and establish a global phase-down of existing oil and gas production. The call included policymakers and CEOs of all oil and gas companies to find solutions, asking them to present credible, comprehensive, and detailed transition plans to reach a net-zero goal, including detailed goals for emission cuts for 2025 and 2030, and efforts to phase out fossil fuels and scale up renewable energy. In addition, the Secretary-General urged that shipping, aviation, steel, cement, aluminum, and agriculture to be aligned with net-zero emissions by 2050. Finally, he called on wealthier nations to deliver climate justice to the most vulnerable of nations by increasing modes of finance and capacities for resilience, adaptation, and loss and damages. He asked banks to provide more grants and concessional loans and for the private sector to fully mobilize private finance. He reminded the wealthier nations about the financial commitments made at the COP in Copenhagen, Paris, and Glasgow.

To protect populations from extreme weather, he called for implementing and expanding early warning systems against natural disasters in the next four years. To help those who would face the front of extreme weather events, he asked for a new loss and damage fund. He ended his press conference by clarifying that the transition must cover the entire economy. Partial pledges won't hit the target and nations must act at warp speed (Guterres 2023).

## James Hansen (1941–)

James Hansen, born in 1941, is an American climatologist who has directed the Program on Climate Science, Awareness, and Solutions at the Earth Institute at Columbia University since 2014. Hansen and his coauthor used the first NASA Goddard Institute for Space Studies (GISS) global temperature analysis of 1981 to analyze surface air temperature from 1880 to 1985. He then established better methods for assessing global temperatures. Hansen found a statistically viable way to combine mid-latitude temperature stations within 1,000 kilometers to assess long-term variability.

At a 1988 senate meeting, he testified that there was a 99 percent certainty that the Earth's average temperature was warmer than ever measured. He added that a clear cause-and-effect relationship existed between global warming, and the increased likelihood of extreme weather. Scientists supported his findings, noting that the rate of temperature change was more significant than at any time in instrument history and concluding that the El Niño of 1988 was not solely responsible for the large temperature anomaly that year.

In 2001, Hansen reported in *Science* that global warming continued and that the increasing temperatures warrant discussions on how to slow it down. By 2006, temperatures were 0.8 °C warmer than the previous century. He concluded that the observed global warming (and the exceptionally high amount of warming in the high latitudes) was an accurate observation of climate change. In 2010, Hansen published a landmark paper titled "Global Surface Temperature Change," describing global temperature analysis.

In recent years, he has been a climate activist, hoping to mitigate the effects of global warming. Hansen represented his granddaughter and "future generations" as a plaintiff in the *Juliana v. the United States* lawsuit, suing the US government and some of its executives for not protecting a stable climate system.

## Stephanie Herring

Stephanie Herring earned her BA from Swarthmore College and her PhD from Yale University. She is a climate scientist at NOAA and editor of the *Bulletin of the American Meteorological Society*. She and her coauthors explain extreme events from a climate perspective. In fact, many of the articles Dr. Herring writes are titled that way, including "Explaining Extreme Events of 2019 from a Climate Perspective." The articles she and her colleagues publish examine extreme weather events, including tropical cyclones, heat waves, high sea surface temperatures in the Pacific and Atlantic Oceans, and human-caused drivers of extreme events. *The Bulletin of the American Meteorological Society* is a unique report that assesses the effect of human-caused climate change on the strength and probability of extreme events for a given year. Herring was honored as one of *Foreign Policy Magazine*'s Top 100 Global Thinkers.

## Hindou Oumarou Ibrahim (1984–)

Hindou Oumarou Ibrahim was born in 1984. She is an environmental activist and geographer who brings awareness to the impact of global climate change on the 250,000 people of the Sahel region who survive on subsistence farming. Oumarou Ibrahim, a member of Chad's Mbororo pastoralists, advocates for the greater inclusion of Indigenous people, their expertise, and their traditions in the global efforts to fight the effects of extreme weather events resulting from climate change. After experiencing first-hand accounts of extreme climate events such as heatwaves and droughts, she began advocating for the Mbororo, who rely on natural resources for survival. In hopes of gaining visibility, she established and is the current president of the Association for Indigenous Women and Peoples of Chad (AFPAT). The tangible effects of climate change on the local environment inspired her to voice the need for emissions regulations and mitigation and adaptation to climate change in the Sahel region.

Oumarou Ibrahim raises concerns for the people of Chad, Cameroon, Niger, and Nigeria, who mainly rely on the water from Lake Chad. She raises awareness that 90 percent of the lake has evaporated or been used for irrigation since the 1960s. In written testimony to the International Organization for Migration, Ibrahim emphasized that the Mbororo and Indigenous communities, like her own, are victims of climate change, emphasizing that her people were forced to abandon their native region for more land that could support them and their livestock. She noted that migrant communities are especially vulnerable to climate change. Ibrahim has also brought awareness to the Mbororo community through her writings in *Quartz* and the World Economic Forum's Agenda.

Ibrahim joined UNESCO and the Indigenous Peoples of Africa Coordinating Committee (IPACC) to create a 3D map of Chad's Sahel desert region using 3D mapping technologies and Indigenous scientific knowledge. The outcome was a tool that could be used by Indigenous peoples, particularly women, to make decisions about mitigating and adapting to extreme events brought on by climate change. Ibrahim represented civil society at the historic Paris Climate Agreement signing on April 22, 2016. At the signing, she stated, "Climate change is adding poverty to poverty every day, forcing many to leave home for a better future."

She is well known as a voice for those who emit the least greenhouse gases but suffer the most from the impacts of climate change. At the 2018 United Nations Climate Change Conference, she debated with the governor of California at the time, Arnold Schwarzenegger's opinion that people can help mitigate climate change by driving electric cars, reducing meat consumption, and other minor everyday actions. Instead, she emphasized the need for governments and policymakers to create change. She also advocates using technology to reduce climate change's impact on desert countries.

She says she aims to spread awareness about the importance of Indigenous knowledge in managing land. In 2019, she delivered a TED talk on the impact of climate change. In her talk, she emphasized the importance of incorporating the opinions of local women in addition to the traditional roles of men in the decision-making process. She pointed out the particular contributions that Mbororo women have made to the location of medicinal and edible plants and the knowledge women have gained from decades of practice. She highlighted the importance of women's ancestral knowledge of weather forecasting, animal behavior, and agricultural techniques. In an interview with BBC for BBC's 100 Women project, Ibrahim stressed that every culture has their science and the Indigenous voice needs to be there.

In 2019, she became one of the seventeen people appointed as advocates of the Sustainable Development Goals by the United Nations. The seventeen nations adopted the United goals in 2015 as a foundation for global sustainability efforts. She is one of the advocates responsible for promoting these goals. Many organizations have recognized her efforts. She received the Pritzker Emerging Environmental Genius Award and was appointed United Nations Sustainable Development Goals Advocate. She is a Member of the United Nations Permanent Forum for Indigenous Issues, a Member of the Indigenous Peoples of Africa Coordinating Committee, a Member of the Advisory Committee to the Secretary-General's 2019 Climate Action Summit, and a Conservation International Senior Indigenous Fellow. She is also the Coordinator of the Association of Peul Women and Autochthonous Peoples of Chad (AFPAT) and co-directed the pavilion of the World Indigenous Peoples' Initiative at the United Nations Framework Convention on Climate Change (UNFCCC) annual meetings COP21, COP22, and COP23.

## Charles Keeling (1928–2005)

Meteorologist and chemist Charles David Keeling is one of the most recognized names in modern climate change and global warming literature. Born in Scranton, Pennsylvania, in 1928, Keeling was an American scientist who developed the first instrument that reliably and consistently measures and records carbon dioxide levels in the tropospheric atmosphere. By measuring the steady increase of carbon dioxide in the atmosphere, he confirmed Svante Arrhenius's proposition that anthropogenic additions of $CO_2$, a greenhouse gas, would lead to global warming.

His science career began at the University of Illinois, followed by a PhD in chemistry from Northwestern University under Malcolm Dole, a polymer chemist with a keen interest in atmospheric chemistry, particularly the geochemistry of carbon and oxygen, the carbon cycle in nature, and air to sea exchange of carbon dioxide. His interests led him to a postdoctoral appointment in geochemistry at the California Institute of Technology. It was at Caltech that he developed the $CO_2$ measuring instrument. From 1961 to 1962 Keeling was a Guggenheim fellow at the Meteorological Institute at the University of Stockholm and, from 1969 to 1970, a guest professor at the Second Physical Institute at the University of Heidelberg as well as the Physical Institute of the University of Bern from 1979 to 1980. After his postdoctoral position, he was convinced by Roger Revelle to join the University of San Diego's Scripps Institution of Oceanography in La Jolla, California. There, he continued his research on CO2 admissions. Throughout his career, he was a member of the American Academy of Arts and Sciences, the American Geophysical Union, the American Association for the Advancement of Science, and of the National Academy of Sciences.

Keeling's first critical theory involved using atmospheric chemistry data and a mass spectrometer to analyze the components of air in forests and rural settings. He concluded that $CO_2$ concentration and isotopic composition in these non-urban settings undergo a regular daily cycle that reflects photosynthesis, respiration, and atmospheric mixing. The data resulted in the Keeling Plot, not to be confused with the Keeling Curve. The Keeling Curve was generated from a plot of $CO_2$ measurements from the Mauna Loa station. Keeling's $CO_2$ instrument initially registered global atmospheric $CO_2$ at 315 parts per million (ppm) in 1958. By 1961, the data showed a steady increase in atmospheric $CO_2$, a trend that has continued since. The resulting graph of measurements is now famously known as the Keeling Curve. The data also clearly showed the seasonal variations as it dips a few parts per million during the Northern Hemisphere's summer when photosynthetic plants use carbon dioxide and a rise in the Northern Hemisphere's winter when deciduous plants cease to photosynthesize. Naomi Oreskes, Professor of History of Science at Harvard University, stated that the Keeling Curve is one of the most critical scientific resources of the

1900s. Many scientists credit the Keeling Curve with initially bringing the world's attention to the increase of $CO_2$ in the atmosphere. The Keeling Curve is considered by many as his most significant contribution to atmospheric science.

The National Science Foundation used Keeling's research in its 1963 warning of rapidly increasing, heat-trapping greenhouse gases. In 1965, President Johnson's Science Advisory Committee also exhibited concerns that extra heat-trapping gases would potentially be dangerous as they would cause the temperature of the Earth to rise. Keeling also published many influential papers during this time, including the 1958 paper titled "The Concentration and Isotopic Abundances of Carbon Dioxide in Rural Areas," the 1965 paper "The Concentration of Atmospheric Carbon Dioxide in Hawaii," and the world renowned 1970 paper "Is Carbon Dioxide from Fossil Fuel Changing Man's Environment?"

The data collection from Keeling's $CO_2$ instrument, which continues to be monitored at Mauna Loa, is the longest continuous record of the world's mid-level tropospheric atmospheric carbon dioxide. Keeling's curve displays the rise in atmospheric concentration of $CO_2$ from 315 parts per million (ppm) in 1958 to 380 (ppm) in 2005, with increases attributed to fossil fuel emissions, which increase atmospheric carbon dioxide. In 2021, the Keeling Curve showed the levels of $CO_2$ reaching 420 ppm for the first time in human history.

Interestingly, his former colleague from the Scripps Oceanography Institution, Roger Revelle, went on to teach at Harvard University in the 1960s, where he displayed the Keeling Curve on the wall in a classroom. His student, Al Gore, who later served as vice president of the United States, saw the curve and became highly interested in the findings. Once elected vice president, Al Gore voiced his concern about the use of fossil fuels and the alarming rate of the accumulation of the gas and its association with the rise in temperatures. In July 1997, Vice President Al Gore presented Charles Keeling with a "special achievement award" for forty years of outstanding scientific research on the atmosphere. Animations of the Keeling Curve have been used countless times to convey the rise in carbon dioxide since 1958. Al Gore famously used the curve in his book and film *An Inconvenient Truth*. His legacy lives on as the curve is still referenced and updated yearly and is now maintained by his son, Ralph Keeling, a climate scientist at the University of San Diego. The Scripps Oceanographic Institute has also animated the curve to help visualize the steady rise in carbon dioxide since 1958. Dr. Keeling was a member of the Association of Meteorology and scientific director of the Central $CO_2$ Calibration Laboratory of the World Meteorological Organization.

Homage to Charles Keeling's contributions is seen throughout the United States. UCSD honors Keeling by developing housing for students of Revelle College at the Keeling apartments, built with environmental and ecological awareness, and by hosting the Charles David Keeling Memorial Lecture Series at the Scripps Institution of Oceanography. The University of Illinois has also hosted Keeling Lectures since 2010. The Keeling Curve is engraved in bronze at the Mauna Loa station and carved into a wall at the National Academy of Sciences in Washington, DC. The Global Warming Mitigation has been awarded the Keeling Curve Prize since 2018, and even plays on the Keeling Curve have been written.

## Thomas Knutson

Thomas R. Knutson is a senior scientist at the US Geophysical Fluid Dynamics Laboratory, a National Oceanic and Atmospheric Administration (NOAA) division. Given global warming scenarios from IPCC models, he aims to clarify the changes to expect in hurricane activity for the later part of the twenty-first century. He also hopes to pinpoint if humans have caused a measurable increase in Atlantic hurricane activity or global tropical cyclone activity. His research focuses on hurricane occurrence, the influence of climate change on hurricane characteristics, the detection of climate change, and the attribution of climate change to extreme weather events.

He is an Associate Editor of the American Meteorological Society's *Journal of Climate*. He has published his findings concerning projections of US landfalling hurricane activity, regional changes in precipitation amounts, and attribution studies concerning climate change and tropical cyclones in prestigious journals, including *Science*, *Proceedings of the National Academy of Sciences*, *Journal of Geophysical Research*, *Journal of Climate*, *Tropical Cyclone Research and Review*, and the *Bulletin of the American Meteorological Society*. He was a lead author on the IPCC's 3rd Assessment Report on *Impacts of Climate Change on Tropical Cyclones in the Typhoon Committee Region*. Knutson was the lead author in Chapter 3: *Detection and Attribution of Climate Change* and Appendix C: *Detection and Attribution Methodologies Overview Climate Science Special Report, Fourth National Climate Assessment* (NCA4).

In 2004, Knutson published the paper "Impact of $CO_2$-Induced Warming on Simulated Hurricane Intensity and Precipitation: Sensitivity to the Choice of Climate Model and Convective Parameterization," suggesting that increases in atmospheric $CO_2$ would lead to more intense hurricanes. Many of Knutson's research papers consider the frequency of intense tropical cyclones, including projections of rainfall amounts. The characteristics are projected for the six major tropical cyclone basins. Knutson projects the characteristics of cyclones if a 2 °C warming of global mean surface temperature was to occur.

According to Knutson's model, projections for a 2 °C anthropogenic global warming include that (1) sea level rise would cause an increase in coastal flooding in the wake of tropical cyclones, (2) an increase in atmospheric moisture content, (3) a 10 to 15 percent increase in rainfall rates averaged within about 100 km of a tropical cyclone, (4) an increase on average of 1 to 10 percent in tropical cyclone intensities globally, with more global storms reaching Category 4 and 5 levels over the twenty-first century, (5) a rapid intensification of tropical cyclones. In general, the projections support a logical increase in the destructive potential of future storms. It is important to note that the frequency of Category 4 and 5 storms is projected to decrease or stay the same as the current frequency.

Hurricane-risk assessment modelers have projected risk scenarios by adjusting past and future global mean surface temperature values. Various risk models have used Knutson's baseline work, including those that estimate how climate change may have already changed tropical cyclones. Knutson concluded that the frequency or intensity increase varies by basin geography. In some basins, the frequency of the most powerful tropical cyclones increases. In contrast, other basins show a decreasing frequency of less intense tropical cyclones.

## Joanne Simpson (1923–2010)

Joanne Simpson was the first woman to gain her PhD in meteorology, which she did in 1949 from the University of Chicago, where she had also completed her undergraduate work. She did postdoctoral work at Dartmouth College. Simpson was a member of the National Academy of Engineering and taught and researched meteorology at numerous universities including the University of Chicago, New York University, Illinois Institute of Technology, Woods Hole Oceanographic Institution, UCLA, the Environmental Satellite Services Administration (ESSA), NOAA, University of Virginia, and the National Aeronautics and Space Administration (NASA) as well as working for the federal government.

Simpson contributed to many areas of atmospheric science; however, her most significant contributions were in tropical meteorology. She contributed significantly to understanding tropical storms, cloud dynamics, weather patterns, and climate. One of the research areas she was particularly interested in was hot towers. A hot tower is a cumulonimbus cloud in the tropics that extends out of the troposphere (the layer of the atmosphere closest to the ground) into the stratosphere. These formations are called "hot" because of the tremendous amount of latent heat released as water vapor condenses and freezes within the cloud. She also studied hurricanes, the trade winds, and air-sea interactions and helped develop the Tropical Rainfall Measuring Mission (TRMM).

She was a passionate advocate for women in the sciences and paved the way for future generations of female scientists. In 1992, she made history by becoming the first woman to receive the Carl-Gustaf Rossby Research Medal, the highest honor in atmospheric science. This was an addition to many other awards she received from the 1950s to the 2020s, including the 1954 Guggenheim Fellowship, the 1962 Melsinger Award from the American Meteorological Society, the 1963 *Los Angeles Times* as "Woman of the Year" in Science, the 1967 Department of Commerce Silver Medal for her work with the Experimental Meteorology Branch, the 1983 AMS Carl-Gustaf Rossby Research Medal, its highest honor, for "outstanding contributions to man's understanding of the structure of the atmosphere," the 2002 prestigious International Meteorological Organization Prize from the World Meteorological Organization. In 2021, the American Meteorological Society established the Joanne Simpson Tropical Meteorology Research Award, granted to researchers who make outstanding contributions to advancing the understanding of the physics and dynamics of the tropical atmosphere.

## Peter Stott

Peter A. Stott is a climate scientist whose expertise includes atmospheric modeling and extreme weather attribution to climate change. He received his undergraduate degree in mathematics from Durham University. He also completed Part III of the Mathematical Tripos at the University of Cambridge. He earned a PhD from Imperial College London for work on atmospheric modeling of the environmental consequences of the Chernobyl disaster. After his PhD, he completed postdoctoral research on the ozone layer at the University of Edinburgh. He is an expert on both natural and anthropogenic causes of climate change. Dr. Stott is part of the Climate Monitoring and Attribution team of the Hadley Centre for Climate Prediction and Research at the Met Office in Exeter, UK.

He is an expert on anthropogenic and natural causes of climate change. He is perhaps best noted in extreme weather and attribution science for his landmark paper published after the extensive heatwaves of 2003 titled "Human Contribution to the European Heatwave of 2003." It was the first quantitative climate change attribution assessment. He and the coauthors found the threshold of record-breaking temperatures during the European heatwave that left over 70,000 people dead in June, July, and August 2003 was doubled by anthropogenic global warming. Stott and his team also found that climate change made the extreme heat event four times as likely. This heat wave is discussed in Chapter 1.

He has also written highly cited papers on the need to assess how climate change will affect various regions of the world. He argues that while knowing the global average temperature rise helps assess the state of the planet, societies need specific temperature data. Scientists, in particular, need climate data sets at daily or even shorter timescales and a resolution of a few kilometers at most to monitor how the weather is changing and to feed local climate models.

Dr. Stott urges the climate community to gather temperature records worldwide and store them in a freely available open database. This would help to answer critical questions about specific regional climate shifts. He also calls for a higher resolution as the only global data sets for temperature are not suited to this purpose. The resolution of the three-monthly average data sets compiled by the NOAA's National Climate Data Center, NASA's Goddard Institute for Space Studies in New York; the UK Met Office Hadley Centre in Exeter, and the University of East Anglia's Climatic Research Unit in Norwich, UK—is only to the hundreds of kilometers. The UK Met Office proposed a higher resolution system and database to the World Meteorological Organization (WMO) Commission on Climatology, which agreed to begin its development.

He was notably a member and lead author of the Intergovernmental Panel on Climate Change Working Group I report, Chapter 9, for the 4th Assessment Report (AR4) published in 2007. He is an editor of the *Journal of Climate*. In addition to many peer-reviewed publications, he is the author of *Hot Air: The Inside Story of the Battle Against Climate Change Denial*. The book is a detailed account of the decades-long battle to deny or downplay the existence of the climate crisis. He was shortlisted for the 2022 RSL Christopher Bland Prize for the book.

## Kevin Trenberth (1944–)

Kevin Edward Trenberth was born in 1944 in Christchurch, New Zealand. He studied at the University of Canterbury, graduating in 1966 with a BSc with the highest honors. His career began with a short stint at the Meteorological Service of New Zealand. He was awarded a research fellowship at the Massachusetts Institute of Technology. He studied under Edward Norton Lorenz and completed his doctoral dissertation, "Dynamic Coupling of the Stratosphere with the Troposphere and Sudden Stratospheric Warmings," in 1972.

Throughout his career, he has contributed significantly to the scientific study of storms and hurricanes under natural and current climate change conditions. He focused much of his earlier work on average temperatures and precipitation. Trenberth's work categorized the intensity, frequency, duration, and type and moisture content of storms. He emphasized that increased water vapor in the atmosphere with higher temperatures leads to greater intensity but less frequency of storms as water vapor is controlled by surface evaporation, not temperature. He voiced his concern to the public about the possibility of more severe storms under global warming scenarios.

In 1998, Trenberth began his renowned work on the augmented characteristics of tropical storms and hurricanes from global warming. In 2004, after four hurricanes made landfall in Florida, Trenberth disputed claims by NOAA officials that the four hurricanes could be attributed to natural climate variability. Trenberth participated in a tele-news conference by Harvard University and cautiously suggested that global warming undoubtedly played some role. This led to extensive criticism by several meteorologists. Trenberth continued researching the subject during the tremendous hurricane season of 2005, including the devastating Hurricane Katrina. A few years later (2007), Trenberth published a piece for the public in *Scientific American* titled "Warmer Oceans, Stronger Hurricanes."

Trenberth's work has also helped clarify certain drops in sea surface temperatures in conjunction with global warming. A long-term warming of sea surface temperatures was due to volcanic eruptions and El Niño. He pointed out that the ocean warming had continued below 700 m depth. In an interview, Trenberth said, "The planet is warming," but "the warmth just isn't being manifested at the surface" (Trenberth 2013.). He pointed to his research, which recorded increased deep ocean heat absorption, particularly after 1998.

As of 2023, Trenberth had published 634 publications plus four videos and many blogs and podcasts. He is also noted to have ranked first in the highly cited papers in environmental science out of 223,246 published from 1996 to 2017. Kevin Trenberth advocates for the scientific evidence of climate change and the urgency to take action. He is known for his statement: "It's real, the problem is cumulative, and we're causing it. Today's blanket of greenhouse gases would disperse only over centuries. Cutting emissions is the most important of all possible responses. We also have to build resilience to the new extremes" ("Kevin Trenberth" 2023).

During his distinguished career, he has been elected to many important positions. From 1989 to 1994, Trenberth led the Tropical Oceans Global Atmosphere program (TOGA). From 1996 to 1999, he was co-chair of the Scientific Steering Group for the Climate Variability and Predictability (CLIVAR) program. He chaired the WCRP Observation and Assimilation Panel from 2004 to 2010 and the Global Energy and Water Exchanges (GEWEX) scientific steering group from 2010 to 2013.

He received many awards and distinguished positions. Trenberth was a Companion of the New Zealand Order of Merit and appointed Distinguished Scholar at NCAR in 2020. In New Zealand, he is an honorary member of the Physics Department at the University of Auckland and an honorary fellow of the Royal Society of New Zealand. In the United States, he is a Fellow of the American Meteorological Society, the American Association for Advancement of Science, and the American Geophysical Union.

The American Meteorological Society honored him with the Jule G. Charney Award in 2000. In 2003, he received the NCAR Distinguished Achievement Award. In 2013, he was awarded the Prince Sultan bin Abdulaziz International Prize for Water, and the Climate Communication Prize from the American Geophysical Union. Trenberth received the 2017 Roger Revelle Medal from the American Geophysical Union for his work on climate change issues. In 2024, he was appointed a Companion of the New Zealand Order of Merit for his contributions to geophysics.

## Organizations

### Berkeley Earth

Berkeley Earth is a US nonprofit organization established in 2013 focusing on environmental data science and analysis. It was founded by Richard and Elizabeth Muller in 2010 when they wanted to investigate legitimate questions of climate skeptics. They organized scientists to reanalyze the Earth's surface temperature record and published their initial findings in 2012.

From 2010 to 2012, Berkeley Earth addressed the five major concerns that global warming skeptics had voiced systematically and objectively. The first four were potential biases from data selection, data adjustment, poor station quality, and the urban heat island effect. Their analysis showed that these issues did not affect the record. The fifth concern they addressed was that there might be an overreliance on large and complex global climate models by the Intergovernmental Panel on Climate Change (IPCC) in attributing the recent temperature increase to anthropogenic forces. Using a 250-year accurate record, they showed that attributions could be made with a simple model that included a volcanic term and an anthropogenic $CO_2$ concentration.

Berkeley Earth houses an open-source database of global air pollution data and global temperature data that they promote as timely, impartial, and verified. The organization uses the data to research the relationships between this data and timely human and environmental issues, such as ocean temperature and acidification. They publish their investigations in peer-reviewed journals, global print and digital media, and through their digital channels.

## Center for Climate and Energy Solutions

The Center for Climate and Energy Solutions (C2ES) was established in 1998 as the Pew Center for Global Climate Change to provide a non-partisan voice to practical climate solutions. The organization is internationally recognized as one of the most influential and reliable think tanks. The organization's leaders and members convene to produce non-biased, research-based climate solutions, which can be given to companies, policymakers, and federal governments. They aim to provide non-biased science to all political spectrums with practical, clean energy-based transformations. C2ES established and continues to organize annual climate leadership conferences and lobby local and federal policymakers to support clean energy development.

One of the solutions supported by C2ES is conducting business to lower greenhouse gas emissions using market-based strategies. A market-based strategy puts a price on emissions. It encourages companies to find cost-effective ways to reduce those emissions or to find a renewable energy alternative. The three primary forms of market-based approaches to reducing carbon in the atmosphere are cap-and-trade, carbon taxing, and renewable portfolio standards (RPS). A government establishes a cap-and-trade approach to set an upper limit on emissions or a "cap." The government sets the cap or the environmental goal, but the market sets the price for reducing emissions. A company can trade allowances with other companies to adhere to their regulations. For example, suppose a company reduces its emissions quickly and cheaply. In that case, it can sell its allowances to a company that finds it more cost-effective to purchase allowances than lessen its own. Companies doing allowance trade encourage their markets to reduce emissions as quickly and effectively as possible.

On the other hand, a carbon tax is when the government sets a price that companies must pay for each ton of greenhouse gas emitted. Under this scenario, markets will determine emissions goals since most businesses will only implement reductions that are cheaper than the tax. Cap-and-trade programs can generate revenue on top of reducing emissions. Both cap-and-trade and carbon tax initiatives can incorporate using carbon offsets or purchasing a reduction, avoidance, destruction, or sequestration of a greenhouse gas to make up for emissions. The RPS approach requires nations or companies to have a portfolio with electricity providers that get a percentage of their electricity from renewable resources. Clean energy can then be traded as Renewable Energy Credits (RECs).

Revenue from cap-and-trade programs or carbon tax efforts can be used to develop clean energy further, reduce taxes, and provide programs that lower energy bills and provide underserved communities with renewable energy options. For example, California has invested more than $2 billion in the reduction of greenhouse gas emissions to efforts that lower energy bills through the use of renewable and energy-efficient technology.

By 2003, C2ES had produced the first bipartisan bill introduced in Congress for climate-based solutions with a market-based consideration. The 2003 bill encouraged a carbon tax or a cap-and-trade program to help companies reduce emissions at the lowest possible cost. The approach also focuses on reducing greenhouse gases and an eventual phase out of leaded gasoline, which will combat acid rain and smog. The idea of a cap-and-trade program on emissions began in the United States in 1990 when Republican President George H. W. Bush launched a market-based solution for sulfur dioxide (SO2) emissions that were causing acid rain. The pollutant was decreased twice as fast as predicted after establishing a cap-and-trade program and at a much smaller amount than the cost of traditional regulation programs implemented in the past.

As of 2023, twelve states are using market-based policies to reduce greenhouse gas emissions (a quarter of the population of the United States). From local governments such as the city of Boulder, Colorado, which was the first in the

United States to implement a carbon tax, to the state of California, which has a cap-and-trade program established in 2013, carbon pricing programs are being used by many places worldwide. European countries have led the way with cap-and-trade programs since 2005. All Canadian provinces have either implemented or are developing pricing plans. The Chinese government is working toward a national strategy, and several Chinese cities and regions have established carbon caps since 2013. Over eighty nations that signed the Paris Agreement of 2015 are planning on developing their market-based solutions to achieve their reduction of greenhouse gases and keep temperatures under the rise of 1.5 °C.

C2ES also hosts webinars where companies such as Bank of America, META, Edison International, and many others can describe ways that they are building climate resilience, defined as the ability to anticipate, prepare for, and respond to acute and ongoing hazardous events—or ways communities can adapt to climate change and recover when severe weather impacts a community where they operate. Corporate leaders discuss how they help protect from the impact of climate change the areas where they have supply chain interests and where they employ hundreds of people. After these types of collaborations hosted by C2ES, publications are compiled and made available for reference by other companies, governments, and policymakers. For example, "An Emerging Blueprint for Companies: Strategies to Advance Local Climate Resilience" was published in September 2022. The document explains the connection between climate change-fueled extreme weather in communities throughout the United States and worldwide part of a global supply chain. The writing promotes strategies that build resilience through corporation resources, funding, and expertise to assist after climate change impacts have disrupted the economy and infrastructure of a location.

Corporations cooperating and collaborating with C2ES have included seven specific ways to advance community climate resilience. The strategies include: (1) reducing company assets and operations risks, (2) developing products and services that enable resilience, (3) supporting the resilience of the company's supply chain, (4) supporting community stakeholders in planning for resilience, (5) empowering employees to volunteer to support community resilience, (6) funding community resilience projects through public-private partnerships, and (7) funding community-led resilience initiatives. The strategies help businesses operate after a disastrous event and maintain economic competitiveness (Brush and Tarvin 2022).

## Climate Central

Climate Central is an organization comprised of independent scientists and communicators who research and report on the changing climate and its impact on various communities. They are the only climate communication group that produces and distributes visual climate change nationally and for localized areas. They are unique in the fact that the organization includes both scientists, journalists, and technologists. They bridge the gap between peer-reviewed research and the general public. The board members range from Princeton University scientists to nonprofit organization leaders focused on communication and diplomacy. The working scientists include Dr. Benjamin Strauss (CEO and chief scientist), cited by the White House and Secretary-General of the United Nations, whose maps titled *Surging Seas* have gathered more than 100 million views. Also on staff is Bernadette Woods Placky, an Emmy Award-winning meteorologist and director of Climate Central's Climate Matters, a program by Climate Central that offers data analyses and graphics to local TV meteorologists and journalists to help them tell local climate change stories.

The organization uses big data and technology to generate scientific conclusions. It creates news stories and visuals for the public. Their primary research is user-friendly maps, tools, data sets, and visuals. The United Nations used their sea level rise visualizations composed by Climate Central at their 2021 UN Climate Conference. They address all levels of the impacts of global warming, including extreme weather. They aim to bridge the gap between scientists and the media. They hope to help individual citizens, communities, businesses, associations, and policymakers understand future outcomes' projections based on various carbon pathways. Their work is used by TV meteorologists, journalists, and other respected voices to reach diverse audiences.

Climate Central organizes its work into what they call Programs. For example, The Sea Level Rise Program focuses on dispersing clear and accurate information about sea level rise and its effects on coastal flooding on a global and local scale. They feature immediate threats and future projections. One of their compelling displays is an interactive map that allows the user to see (by selecting on a sliding scale) global warming scenarios of 1.5 °C to 3.0 °C and the changes to coastlines on a map by 2100 and beyond. They also feature images of 190 locations worldwide with projected sea level after a warming of 1.5 °C and sea level after a warming of 3 °C within a 200 to 2,000-year time frame.

The images show flooding levels and their extent. Climate Central offers links to peer-reviewed research papers that supply the data for the images. The images visualize sea levels and extreme weather's impacts on these communities.

Perhaps most pertinent to the scope of this book is Climate Central's Climate Matters Program. Climate Matters is the division of Climate Central that enables data-driven climate reporting for media networks. The program aims to communicate global climate change to local audiences in ways that "matter." Climate Matters produces bilingual (English and Spanish) climate news each week specific to their area (245+ cities). Communicators, such as local meteorologists and educators, can use data analyzed by Climate Matters and localized graphics and science videos on current weather events. Climate Central's Partnership Journalism Program collaborates with newsrooms and scientists. For example, KUNR and Climate Central did a story titled "Fire Extremes Make Prescribed Burns Riskier, but More Essential Than Ever." Finally, the Climate Shift Index Program reveals the influence of climate change on local weather. The index reveals the "shift" in the occurrence frequency of a particular weather event due to climate change. It shows the likelihood of the temperature or other weather events currently experienced by a population by giving the event a Climate Shift Index (CSI) value ranging from −5 to +5. Each level above 2 is a multiple of how frequently a particular temperature will occur due to global warming. A level of 5 on the CSI scale for a Salt Lake City heat wave, for example, means that it occurs five times more frequently and would be highly unlikely to experience the event without the current warming. The more positive and the greater the CSI value, the more likely the event will occur. Negative CSI values denote the event is less likely due to global warming. Meteorologists and educators can display Climate Central's CSI graphs and images to show the public a clear connection between the weather events they recently experienced at their particular location (wildfires, floods, hurricanes, etc.) and global warming.

## Climate Transparency

Climate Transparency is a global partnership that produces annual reports assessing the climate performance of G20 countries. These reports comprehensively analyze a country's progress toward climate goals, including emissions reductions, renewable energy deployment, and policy effectiveness. Policymakers can rely on these reports to evaluate the effectiveness of their climate policies and identify areas for improvement. Business leaders can use the information to inform investment decisions, anticipate regulatory changes, and align corporate strategies with climate targets. Additionally, the general public gains valuable insights into their country's climate efforts, fostering accountability, and encouraging civic engagement.

## Columbia Climate School

The Columbia Climate School, established in 2020, is the first climate school in the United States. It was established at Columbia University to attract bright students whose goals are future careers that solve the climate crisis. Incidentally, the term "Global Warming" originated at Columbia. Students are given the opportunity to work with award-winning faculty who use an interdisciplinary approach in their research and teaching to address and resolve the impacts of the current climate and prepare them to integrate solutions for society. The Columbia Climate School also conducts innovative climate change research incorporating more than twenty-five transdisciplinary research centers, including those that provide the components of finance, sustainability, and engineering, to name a few. The Columbia Climate School has over five hundred PhD research students working with nearly five hundred funding partners. The institution has been granted over one billion dollars in funding since 2013. The research provides evidence-based conclusions on the state of the planet. It provides policymakers, nonprofits, and communities with the best information for mitigating and adapting to climate change and extreme weather events.

The Columbia Climate School also provides breaking climate-related news stories for the public on pressing topics and recent innovations worldwide. The news stories are written at a level that relays the most important discoveries to a general audience. Many of their articles provide new findings, attribution of extreme events to global warming, and societal issues resulting from extreme weather events.

## Global Change Research Program

Although climate change is often used as a debate topic for politically heated discussions, the truth is that since the late twentieth century, the US government has been researching climate change with genuine concern about what global warming will do to the regions of the United States. As the government and the public experience first-hand the longer and hotter summers, extended heatwaves, freezing polar vortex days in the Midwest, drought in the American West, and hurricanes at all latitudes of the coast, they become more curious about climate change and the science that has brought such changes. Advances in technology and computer science have made climate change science increasingly confident and statistically significant. It is now proven without a scientific doubt that climate change is happening and is caused by the human-produced excess of greenhouse gases in our atmosphere. Scientists are still refining projections for the future. Still, current observations are conclusive that the last fifty years of warming is primarily due to burning fossil fuels, clearing forests, and some agricultural practices. Scientific evidence has proven that without the increase of greenhouse gases produced by human activity, the planet would have actually cooled slightly over the past fifty years.

The Global Change Research Act of 1990 required that the US Global Change and Research Program (USGCRP) compose a report to be delivered to Congress and the President at least every four years. The publication aims to convey the effects of global climate change on the natural environment, agriculture, energy production, land and water resources, transportation, health, welfare of communities and individuals, social systems, and biological diversity. The report is the National Climate Assessment and is currently in its fourth iteration. An important feature of the report is the regional breakdown of climate change. It provides summaries for assessing the Northeast, the Southeast, the Midwest, the Great Plains, the Southwest, the Northwest, Alaska, Hawai'i, rural communities, and the coastal regions.

The National Climate Assessment summarizes the impacts of climate change, including severe weather, on the United States now and into the future using climate projections derived from climate models. The reports help Americans assess observations and research to understand what is happening to our weather as the globe has warmed. More than 300 experts, including people from federal, state, and local governments, Indigenous communities, national laboratories, universities, and corporations, volunteered to produce the report guided by a team of sixty Federal Advisory Committee members published the report after a peer review process (reviewed by the public, experts, federal agencies, and a panel of the National Academy of Sciences). Each region of the United States (over forty cities) hosted workshops. The National Oceanic and Atmospheric Administration (NOAA) was the lead agency in preparing the fourth and latest report.

The Fourth National Climate Assessment (NCA4) is divided into two volumes. The first volume (VI) of the Climate Science Special Report (CSSR) was published in 2017. This report focuses on the physical science of climate change and how climate change is affecting the Earth's system. It looks at trends from anthropogenic and natural climate forcers (physical science properties that influence climate change). It also provides an assessment of the latest findings in climate science. It discusses the confidence level, limitations, and uncertainties in those findings, such as projected changes in temperature, precipitation patterns, and sea level rise, based on Representative Concentration Pathways or RCPs, which were discussed in Chapter 1 of this book and which are used by many climate scientists and policymakers. The second volume (VII) focuses on social science issues that affect human welfare, societies, and the environment within the scope of climate change for ten regions and eighteen national topics.

## Iberdrola

Iberdrola is a clean energy group committed to advancing climate change solutions. It emphasizes the importance of global organizations recognizing the most vulnerable countries and agreeing to help them in their climate resilience. Despite the various views at COP27 by developed nations on which exact countries are the most susceptible, Iberdrola set to the task of looking at which nations were undeniably most at risk and would suffer the greatest from climate disasters and extreme weather events. Iberdrola gathers information from many agencies and develops a list of the most vulnerable countries.

## Intergovernmental Panel on Climate Change

The Intergovernmental Panel on Climate Change (IPCC) is an international organization established by the World Meteorological Organization (WMO) and the United Nations Environment Program (UNEP) in 1988. The IPCC was founded at a time when many nations around the world began to understand the severity of the climate problem. It is considered the leading authority on climate change and all aspects related to global warming. The organization aims to review and communicate the most recent research on anthropogenic (human-caused) global warming, resulting climatic changes (including extreme weather), and their impacts on society. Leading climate scientists and all 195 member governments endorse its findings.

Governments and organizations select scientists to serve for an assessment cycle of six to seven years. During that time, the experts prepare assessment reports for policymakers and the general public about various aspects of the planet associated with global warming. The IPCC's scientists compose three working groups. Working Group 1 (WGI) focuses on the physical science aspects of the climate system and climate change. WGI assesses the state of the atmosphere, including greenhouse gases and aerosols, air temperature, and rainfall fluctuations. The group also analyzes changes in historical and paleoclimate variations in land and air temperatures and the state of the oceans (temperatures, sea level rise). Their tasks include reviewing publications that run climate models to project variations to the climate system based on greenhouse gas emissions. They also synthesize the findings from research that analyzed satellite data and other forms of data to try and find causes and attributions to climate change. Working Group 2 (WGII) comprises social scientists who assess the vulnerability of various populations at various socioeconomic levels to climate change. This second working group also advises on adaptations to increasing global temperatures. One of their essential tasks is to find sustainable options for water resources, energy, food, business, and human health while maintaining the ecological health of the forests, oceans, and land. Working Group III (WGIII) assesses mitigation strategies to reduce the amount of global warming from greenhouse gas emissions. They look at promoting innovations within the energy, forestry, transportation, waste management, construction, and agriculture sectors that diminish the amount of greenhouse gases in the atmosphere or remove them from the atmosphere.

After thousands of hours of review, collaboration, and deliberation on the best way to convey the importance of the working groups collaborating to publish annual reports, the first, completed in 1990, was titled "The Assessment Report 1" (AR1). This report established the United Nations Framework Convention on Climate Change (UNFCCC). The UNFCCC is an international treaty among countries to deal with the problem of human-produced overabundance of greenhouse gases in the atmosphere and whose goal is to find ways to avoid "dangerous human interference with the climate system," including those that affect the atmosphere, the hydrosphere, and the biosphere. The treaty was signed in 1992 by 154 nations at the United Nations Conference on Environment and Development (UNCED), informally known as the Earth Summit of Rio de Janeiro or the Rio Summit. The treaty began on March 21, 1994, and calls for ongoing scientific research by scientists with expertise in Earth, biological, and social sciences focusing on climate change and societies. The framework also establishes regular meetings known as the Conference of Parties or COPs, negotiations, and future policy agreements designed to allow animals and plants to adapt, ensure food production, and enable sustainable economic growth sustainably.

One of the UNFCCC's most recognized efforts was the Kyoto Protocol. The Protocol set guidelines for industrialized nations (or nations hoping to become industrialized) to reduce greenhouse gas emissions under agreed-upon individual targets. The Protocol was adopted in 1997, but due to a complex ratification process, it took effect on February 16, 2005. Initially, the UNFCCC asked countries to devise policies and continue efforts to mitigate climate change and periodically report those. The Kyoto Protocol bound thirty-seven industrialized countries plus the European Union to their pledged reduction. Higher efforts are required for those who emit more. Overall, the targets set an average goal of 5 percent emission reduction (compared to 1990 levels) over the preliminary commitment period from 2008 to 2012.

The Kyoto Protocol also established flexible market options based on the trade of emissions permits. Under the Protocol, countries were encouraged to meet their targets primarily through national efforts, especially those that reduce greenhouse gas emissions or develop new technologies to reduce emissions. In addition, the Protocol offers nations three other market-based solutions, including international emissions trading, clean development mechanisms, and joint implications. Parties submit reports annually to ensure the success of the Protocol. The reports are rigorously reviewed and verified to ensure honesty.

The United Nations Climate Change Secretariat in Bonn, Germany, records and tracks emissions, emission reductions, and market-based solutions. The secretariat keeps an international transaction spreadsheet to verify that transactions were consistent with the laws of the Kyoto Protocol. A compliance system was established to ensure nations falling short of their proposals can get on track to meet their commitments so that they may receive the aid needed. Parties of the Kyoto Protocol benefit from the Adaptation Fund established to finance adaptation projects and programs in developing countries. The funding is acquired from climate-friendly projects and international emissions trading. The Doha Amendment of 2012 extended the commitments set by the Kyoto Protocol from 2013 to 2020, adding a revised list of greenhouse gases.

The 2015 United Nations Climate Change Conference (COP 21) in Paris, France, was the 21st annual meeting of the Conference of the Parties (COP) established by the UNFCCC. The meeting's most significant accomplishment was the establishment of the Paris Agreement. The Agreement is a binding international treaty to reduce greenhouse gas emissions to keep temperatures from increasing by 2 °C (preferably 1.5 °C) above preindustrial levels and reduce climate change's effects. One hundred and ninety-five parties adopted the Agreement. This was the first time that an international binding agreement focused on the issue of climate change. The Paris Agreement required fifty-five nations representing at least 55 percent of global emissions to join formally. That requirement was met on October 5, 2016, and the Agreement was enacted thirty days later on November 4, 2016. Since 2016, nearly every nation (197 countries) has signed the pledge and vowed to reduce their emissions.

The Special Report on Global Warming of 1.5 °C in 2018 by the IPCC related global greenhouse gas emission pathways to future scenarios and had startling news for the public. The key findings were that it was still possible to keep global temperatures from rising beyond 1.5 °C (2.7 °F), but that it would take a global effort and "rapid, far-reaching and unprecedented changes in all aspects of society." The report also clarified that to achieve the 1.5 °C target, $CO_2$ emissions must decline by 45 percent (relative to 2010 levels) by 2030, and net-zero emissions must be reached by 2050. The previous pledge of the countries following the Paris Agreement would still result in a global rise in temperatures of 3.1 °C to 3.7 °C by 2100. That rise in temperatures would increase risks to human health, raise sea levels, diminish Arctic Sea ice, bleach coral, threaten ecosystems, and increase the chance for extreme weather.

The associated risks of a rise from 1.5 °C to 2 °C could trigger increases in average temperature in most land and ocean regions, prevalent hot extremes, heavy precipitation in several regions, and the loss of moisture in some areas. For example, sweltering days in mid-latitude regions could warm by up to 3 °C at a global warming of 1.5 °C and 4 °C at 2 °C, and extreme cold nights in high latitudes warm by up to nearly 4.5 °C with a 1.5 °C and about 6 °C with an increase of 2 °C. The number of hot days will increase with the highest increases in the tropics.

High-level ministers urged that climate action be upscaled to reach the Paris Agreement goals at the Conference of Parties 27 (COP27) held in Egypt in November 2022. They stressed the urgency highlighted in the IPCC report published the preceding year that called for global emissions of greenhouse gases to peak before 2025. The latest IPCC report emphasizes that anthropogenic greenhouse gases are responsible for approximately 1.1 °C of warming from 1850 to 1900. Using improved observational data sets, scientists were better able to assess historical warming. Advances in the scientific understanding of the climate system's response to human-caused greenhouse gas emissions were also used to project future likely scenarios. The IPCC found that, on average, in the next two decades, global temperature is expected to rise by 1.5 °C.

Notable contributions and accolades for the work of the IPCC include the panel sharing the 2007 Nobel Prize with Al Gore for their combined contributions. Formal approval has been given to 190 nations for their proposed pathway to achieve the Paris Agreement goals. Iran, Türkiye, and Iraq are the only large countries emitting greenhouse gases that have not signed the Agreement since 2024. The 5th Assessment Report (AR5) was instrumental in the well-known Paris Agreement of 2015. The AR6 report was accompanied by three reports pertinent to extreme weather and climate change.

## Loss and Damage Collaboration

The Loss and Damage Collaboration (LDC) is a London, England-based group of researchers, activists, and policymakers from both the Northern and Southern Hemisphere whose aim is to ensure that populations in vulnerable countries have resources to mitigate the impacts of climate-related loss and damage especially those caused by extreme weather events. The LDC plans to work with developed countries to advocate for vulnerable populations and encourage channeling resources to those in need. The group also conducts training for youth around the world. It creates digital storytelling to communicate how young people are impacted by climate change.

The organization's youth division is the Loss and Damage Youth Coalition (LDYC). It was established in 2020 when young climate activists worldwide felt it essential to create awareness and promote aid for climate-induced loss and damage. Youth activists engage, lead, and collaborate to address loss and damage due to climate change.

## Met Office Hadley Center

The Met Office Hadley Center, established in 1990 in the UK, provides climate research and data for decision-makers, researchers, and the general public. Its research incorporates monitoring and reporting climatic changes and their causes. Many of its research projects focus on increasing the understanding of the physical, chemical, and biological processes within the climate system.

Their scientists have worked with international researchers from over 1,700 institutions from 138 countries. Two lead authors, two coordinating lead authors, and one review editor from the organization worked on the Intergovernmental Panel on Climate Change's sixth assessment Report (AR6). The Met Office Hadley Center has also produced more than 2,200 peer-reviewed publications in scientific journals across 200 journals and have been cited by 200,000 researchers.

They also develop services that help with mitigation and adaptation to climate change. The services include the monitoring of global and national changes to the climate. Researchers at the center have also developed state-of-the-art computer models to analyze the current state of the planet. Using models helps to find the causes of recent changes in climate and extreme weather events. They also project future extreme weather possibilities using various emissions scenarios. Their research and services provide for the needs of the UK government by providing policy-relevant scientific evidence and advice.

## NASA's Global Climate Change, Vital Signs of the Planet

NASA's Global Climate Change, Vital Signs of the Planet website includes a yearly reference to the changes occurring in Earth's environment since preindustrial times, including the concentration of $CO_2$ and methane ($CH_4$). It also offers a quick reference to the change in sea level since 1993. The billions of metric tons of ice sheets are listed annually, and the percentage of Arctic Sea ice has been at its minimum extent since 1979. The sea level rise has been listed in inches since 1993. Ocean warming has been listed since 1955. This offers a quick view of the physical components that affect extreme weather events. NASA also provides links to climate change research by NASA scientists, ways to visualize and explore changes in the physical components of global warming through time, and many facts, current news, solutions, and links to resources. Users can monitor Earth's climate with satellite photos, easy-to-follow graphs, and articles.

## National Oceanic and Atmospheric Association

The National Oceanic and Atmospheric Association (NOAA) is an internationally respected organization based in Washington, DC, within the United States Department of Commerce. The agency is one of the leading agencies tasked with monitoring the state of the planet and national environmental conditions.

The agency is highly influential in shaping space, climate, and weather policies. President Richard Nixon established the agency in 1970 after seeing a need for the "better protection of life and property from natural hazards ... for a better understanding of the total environment ... [and] for exploration and development leading to the intelligent use of our marine resources" (Nixon 1970).

Since its inception, NOAA has committed to monitoring many aspects of our planet and recording, archiving, and communicating changes in the Earth's systems. NOAA's scientists use the latest technology to collect data and discern changes in our natural world. NOAA's Satellite and Information Service sector provides public data and images from the observed global environment. NOAA researchers have state-of-the-art satellites, laboratories, research ships, science centers, and many honored scientists and workers who support research programs. Their resources help policymakers and the public understand a dynamic planet. They also strive to communicate environmental stewardship and protection.

NOAA's administrators have addressed the ever-increasing risks to life, property, economies, and ecosystems from the alarming rate of global warming. The association has declared its commitment to addressing the climate crisis as a top priority. NOAA has established a reputation as the United States' authoritative source for rigorous science, data collection, observation, and forecasting and is a leading provider of climate change products and services. The association provides climate data and services to underrepresented nations at risk for hazardous situations due to extreme weather. Evolving measures include collaborating with coastal communities to adapt to sea level rise, providing early warning signs for drought to agriculturalists and tribal and Indigenous communities, aiding emergency operations before, during, and after extreme weather events, and tracking the trends for sea level rise, global temperatures, and the paths of hurricanes, tornadoes, floods, and forest fires.

As a commitment to staying at the forefront of extreme weather events, NOAA collaborates with physical and social scientists to improve awareness of extreme heat and cold forecasts, significantly helping support vulnerable and underserved communities. They aim to provide a context for heat waves and cold waves. Using data from the Census Bureau, various agencies, and the public, NOAA will compare the magnitude of current weather events in a historical context. The agency provides websites, animations, maps, satellite imagery, and other resources at the click of a mouse.

Various branches of NOAA provide services directly related to extreme weather and climate change. For example, the National Weather Service (NWS) provides weather forecasts and warnings in hazardous weather to protect the public and publish general information. The National Center for Environmental Information (NCEI) houses over sixty petabytes of environmental data. The data is available to help decision-makers address specific challenges concerning extreme weather events. They boast that data from the bottom of the oceans to the surface of the sun is available to the public. NCEI uses the data to create scientific products and services that can be valuable resources for agriculture, transportation, the retail world, legal services, energy and utilities, coastal managers, tourism, and many other sectors. The National Climate Data Center, also a division of NOAA, provides climate information, including The Global and Historical Climate, containing worldwide historical temperatures and precision data. Other divisions of NOAA, all of which deal with the effects of global warming and extreme weather directly or indirectly, include the National Environmental Satellite, Data, and Information Service, the National Marine Fisheries Service, the National Ocean Service, the Office of Marine and Aviation Operations, and the Office of Oceanic and Atmospheric Research.

NOAA is committed to a five-year plan to advance its capabilities to assess, identify, and predict extreme weather events. The NOAA FY22-26 Strategic Plan provides the association with significant priority areas, including (1) to build a climate-ready nation by establishing NOAA as the primary federal authoritative provider of climate information for the government's response to the climate crisis, (2) to integrate equity into core operations, and (3) to promote economic development while maintaining environmental stewardship while focusing on supporting the New Blue Economy (the sustainable use of ocean resources for economic growth). The association's focus in the next five years includes, but is not limited to, the following. The aim is to improve forecast models and continue to develop and implement NOAA's complete Earth system—prediction capabilities for extreme weather events. With the understanding of the high risk of flood events in the future, NOAA is committed to offering emergency managers real-time forecasting and flood inundation mapping (FIM) capability at the street level nationwide to assist in planning warning and relief efforts. They plan to develop regional drought early warnings and forecasts for communities and economies nationwide, covering weather-to-climate (days to decades) timescales. To achieve this goal, they plan to expand and enhance the Droughts Interagency National Integrated Drought Information System

(NIDIS). The association also aims to improve its monitoring of wildfires by integrating observations, modeling, meteorology, artificial intelligence, and machine learning algorithms to vastly improve the early detection of wildfires from satellite data.

## United Nations Environment Program

The United Nations Environment Program (UNEP) Adaptation Gap Report (AGR) series provides an annual science-based assessment of global adaptation planning, financing, and implementation progress. Since 2014, UNEP has consistently produced the AGR to provide insights to inform climate negotiations among United Nations Member States. The 2023 report's title, "Underfinanced. Underprepared. Inadequate Investment and Planning on Climate Adaptation Leaves World Exposed," summarized its content. The UNEP report highlighted the significant implications of the failure to effectively address the severe impacts of climate change on the most vulnerable populations. Much of the funding is needed to develop food and agricultural systems resilient to climate change, develop resilient medical services to treat health impacts arising from climate change, and reduce climate-related morbidity. Technological advances are also needed, such as renewable energy sources or low-carbon transportation and enhancing the resilience of water sectors to physical climate effects.

The UNEP reported that developing countries need up to $387 billion a year to adapt to climate change, an increase of $47 billion since the last assessment. The aid is expected to come from developed nations. Still, international public cash is evaporating, and the effects of global warming are becoming more disastrous. The findings arrived just ahead of the COP28 climate summit when the question of international funding for countries coping with extreme weather was high on the agenda. The UN found that the need for "adaptation" aid was 10 to 18 times more than international public finance flows, accounting for a gap of about $360 billion. UNEP also reported findings detailing that the fifty-five most vulnerable nations experienced losses and damages of more than $500 billion from the early 2000s to the 2020s, a figure they predict would increase as the effects of climate change worsen.

As mentioned several times in this book, scientists from the international community agree that extreme weather events will become more prevalent and induce more damage with every fraction of a degree of global warming. UNEP suggested various ways of increasing financing for adaptation, including reformation of the international financial architecture to overhaul the World Bank and other multilateral development banks. Attendants of COP 28 were urged to increase their commitment to helping buffer the devastations faced by many low-income countries and disadvantaged groups from damaging climate impacts or, at the very least, live up to the responsibilities that wealthy nations set at the COP26 in Glasgow when countries pledged to provide nearly $40 billion annually in adaptation finance support by 2025.

## World Meteorological Organization

The World Meteorological Organization (WMO) is a United Nations agency focused on all aspects of weather and climate. It evolved from the 1873 Vienna International Meteorological Congress and later from the International Meteorological Organization (IMO). The organization was established by the ratification of the WMO Convention in 1950. It provides valuable worldwide information on weather and climate.

The WMO can provide global weather information through data collected from 10,000 manned and automatic surface weather stations, 1,000 upper-air stations, 7,000 ships, 1,100 moored and drifting buoys, hundreds of weather radars, and 3,000 specially equipped commercial aircraft gathering all forms of weather data and atmospheric data from land and ocean every day. In addition, the WMO operates thirty meteorological and 200 research satellites for meteorological, hydrological, and geophysical observations. Powerful computers at the WMO process the data collected from their satellites and buoys. The data is used in numerical models based on the laws of science to produce weather, climate, and water-related forecasts. After compiling the estimates, the WMO Information System (WIS) provides an automated transfer of observed weather and climate. WMO data from its main center and all other centers under its jurisdiction are free. The WMO is responsible for updating and strengthening how it communicates the information worldwide to satisfy the need for weather, climate, and water monitoring information as all sectors of society face issues from the effects of climate change and the increasing frequency and impact of extreme weather.

Decision-makers use the warnings of severe weather and air quality reports to prepare for short-term inclement weather and long-term climate events. The data provided by the WMO is used to save lives and prevent property loss. WMO has committed to helping nations with climate change mitigation and adaptation and sustainable development. To accomplish this, the WMO has established an Earth Systems observation network, a data management center, a quality control center for the uniformity of practices and procedures defined by the WMO Instruments and Methods of Observation Program (IMOP), the development of science to reduce disaster risk and climate change adaptations, and a commitment to the fields of aviation, water resource management, agriculture, health, and energy. The WMO also coordinates research and training for professionals in meteorology and related fields.

WMO programs provide its members with the latest standards and data in severe weather and climate change. For example, the agricultural meteorology program (AgM) supports food and agricultural production and activities by helping to meet the needs of farmers, herders, and fishermen to sustainably develop farming systems, improve production and quality of crops, reduce monetary losses, decrease agricultural business costs, and risks, increase efficiency in the use of water, conserve natural resources, and decrease pollution from chemicals.

The Associated Program on Flood Management (APFM) assists in mitigating the most common natural hazards with the most considerable impacts on society. The loss of life has decreased due to improved early warning systems; however, much work appears to be needed to diminish damage to land and property. The WMO reports that insufficient prevention, economic growth, and a lack of flood-sensitive land use planning have increased flood risks. APFM aims to minimize fatalities while maximizing the efficient use of floodplains. The APFM provides an online demand-driven technical support site where users can request customized technical support or explore various available flood management ideas. The Disaster Risk Assessment Program (DRR) assists members in establishing plans to protect populations from natural hazards. Economically feasible and sustainable proposals are provided to members so that they may adapt to our changing climate. The main components of the DRR program encompass broad support for developing early warning systems and risk assessment, an invaluable service in the face of increasing extreme weather events.

Other WMO programs offer the most current information for mitigating life and property risk. The Global Atmospheric Watch (GAW) Program provides information and services on atmospheric composition to the public, policymakers, and decision-makers regarding the number of atmospheric greenhouse gases, ultraviolet radiation in the stratosphere, and urban air pollution. The global atmospheric data provided by GAW is used during international conferences. The Global Climate Observing System (GCOS) is a co-sponsored program by the World Meteorological Organization, the Intergovernmental Oceanographic Commission of UNESCO (IOC-UNESCO), the United Nations Environment Program (UNEP), and the International Science Council (ISC), which regularly assesses the status of global climate observations and produces guidance for the improvement of physical, chemical and biological methods of observations that ultimately contribute to solving challenges in climate research. Finally, the Global Data-processing and Forecasting System (GDFS) provides high-quality products and services for weather, climate, water, and environment in all timescales.

## World Weather Attribution

The World Weather Attribution (WWA) was founded in 2015 by Dr. Friederike Otto, Dr. Geert Jan van Oldenborgh, and the nonprofit Climate Central (featured in its own profile). The WWA works with scientists worldwide to statistically quantify the influence of global warming on extreme weather events. They are unique in launching research efforts immediately following an occurrence such as a heatwave, a hurricane, a wildfire, or a mid-latitude cyclone. They use ground observation data and computer modeling to accomplish this task. The results are posted on their web page after the model outputs are analyzed. The availability of the information raises awareness about the need to curtail emissions to avoid future similar events. It also highlights the need to support vulnerable communities with resources, infrastructure development, and financial aid.

For example, the headlines on their front page during the early summer months of 2024 included, but were not limited to, the following: "Climate Change, El Niño and Infrastructure Failures behind Massive Floods in Southern Brazil"; "Urban Planning at the Heart of Increasingly Severe East African Flood Impacts in a Warming World"; "Autumn and Winter Storm Rainfall in the UK and Ireland Was Made about 20% Heavier by Human-caused Climate Change"; "Climate Change Made the Deadly Heatwaves That Hit Millions of Highly Vulnerable People across Asia More Frequent and Extreme"; and "Low Water Levels in Panama Canal Due to Increasing Demand Exacerbated by El Niño Event." Since 2015, WWA has performed more than sixty attribution studies on extreme weather events worldwide. The methods used have been refined as technology advances. The findings have been published in peer-reviewed scientific literature.

## Yale Program for Climate Change Communications

The Yale Program for Climate Change Communications conducts scientific research on public climate change knowledge, attitudes, policy preferences, behavior, and the psychological, cultural, and current influence of political factors. It aims to engage the public in the science of climate change and solutions, forming partnerships with governments, media organizations, companies, and communities. The program produces a radio program, Yale Climate Connections, which airs daily. On their website, users can find the latest news and media, research articles, and interactive tools. They can also use pulldown menus to create maps or charts showing survey results for a variety of topics, including by county, state, or the entire United States.

Anthony Leiserowitz and Sarah Lowe are the directors of the Yale Program on Climate Change Communication, a Senior Research Scientist at Yale School of the Environment, and a clinical psychologist and Associate Professor in the Department of Social and Behavioral Sciences at Yale School of Public Health, respectively. They both study climate anxiety. According to Lowe, climate anxiety is distress about climate change and its impact on the environment and human existence. It can produce intrusive thoughts or feelings of distress about future disasters or the long-term future of human existence, including one's future family descendants. Clinically, a physiological response manifests as the heart racing and shortness of breath. The anxiety can also impact a person's behavioral component when climate anxiety impedes social relationships or a person's ability to function at work or school.

Leiserowitz has been surveying Americans on their level of concern, worry, and anxiety about climate change for more than twenty years, finding a substantial number who now say they cannot stop fretting about the risks posed by global warming. He distinguishes in his writings the difference between concern, worry, and anxiety. The concern is broad, while worry, he points out, can lead to investigation, action, and mitigation. In contrast, anxiety becomes overwhelming and debilitating, keeping someone from living their life. Worry has steadily increased within the United States from 2000 to 2020. About 27 percent of people in the United States report being apprehensive about climate change.

Their studies (as of 2023) found that around 10 percent of people surveyed report feeling concerned about the global rise in temperatures. Nine percent noted being unable to stop or control worry about global warming, and 7 percent lost some interest or pleasure in doing things because of global warming. Research also indicated that climate anxiety is highly correlated with generalized anxiety. Lowe found statistical evidence that among young adults, in particular, those who were not engaged in group activities to solve the issue had higher climate anxiety. A large body of literature confirms that social support encourages mental well-being. People who engaged in collective action had less anxiety or worry, particularly if their actions had an impact. Some research, though, shows that people who engage in activism can be at risk of experiencing the stressors of burnout. Studies also showed that longer-term collaborative efforts, where people can interact, establish relationships, and see success from their efforts, such as policy changes, are theoretically most beneficial.

When looking at socioeconomic groups and developed nations versus developing nations of people who have heard of climate change (on a global scale, there are still about two billion people who have never heard of climate change), people in developing countries tend to be much more worried than people in the developed world. They know that they do not have the resources to adapt to extreme temperatures, storms, and droughts.

## Further Reading

Brush, L., and S. Tarvin. 2022. "Major Companies Invest in Building Local Climate Resilience." Center for Climate and Energy Solutions, October 11, 2022. Accessed October 15, 2024. https://www.c2es.org/2022/10/major-companies-invest-in-building-local-climate-resilience.

Clark, C. 2005. "Obituary Notice: Climate Science Pioneer: Charles David Keeling." Scripps Institution of Oceanography, June 21, 2005. Accessed October 15, 2024. https://scripps.ucsd.edu/news/obituary-notice-climate-science-pioneer-charles-david-keeling.

Guterres, A. 2023. "Secretary-General Video Message for Press Conference to Launch the Synthesis Report of the Intergovernmental Panel on Climate Change." United Nations in Jordan, March 20, 2023. Accessed October 15, 2024. https://jordan.un.org/en/223957-secretary-general-video-message-press-conference-launch-synthesis-report-intergovernmental.

"Hindou Oumarou Ibrahim." n.d. Akina Mama wa Afrika. Accessed July 25, 2024. https://www.akinamamawaafrika.org/hindou-oumarou-ibrahim.

IPCC. 2007. "Chapter 4: Observations: Changes in Snow, Ice and Frozen Ground." IPCC Fourth Assessment Report: Climate Change 2007. Accessed October 15, 2024. https://archive.ipcc.ch/publications_and_data/ar4/wg1/en/CH4.html.

IPCC. 2021. "IPCC Sixth Assessment Report: Climate Change 2021." August 12, 2021. Accessed October 15, 2024. https://kabcnt.org.au/ipcc-sixth-assessment-report-climate-change-2021.

Jewson, S. 2021. "Conversion of the Knutson et al. Tropical Cyclone Climate Change Projections to Risk Model Baselines." *Journal of Applied Meteorology and Climatology* 60, no. 11. https://doi.org/10.1175/jamc-d-21-0102.1.

"Kevin Trenberth: Communicating Climate Science with Urgency." 2023. The University of Auckland, August 1, 2023. Accessed October 15, 2024. https://www.auckland.ac.nz/en/news/2023/08/01/kevin-trenberth-climate-change-scientist.html.

The Kyoto Protocol. n.d. Republic of Rwanda. Accessed July 25, 2024. https://climatechange.gov.rw/index.php?id=82.

Lieu, A. n.d. "'Fingerprints' of Climate Change Marking Increasingly Extreme Weather Events, Research Suggests." Fox 5 Atlanta. Accessed July 25, 2024. https://www.fox5atlanta.com/news/fingerprints-of-climate-change-marking-increasingly-extreme-weather-events-research-suggests.

Loss and Damage Collaboration. n.d. *Unitarian Universalist Service Committee*. Accessed July 25, 2024. https://www.uusc.org/initiatives/loss-and-damage-collaboration.

"Market-Based Strategies." n.d. *Center for Climate and Energy Solutions*. Accessed July 25, 2024. https://www.c2es.org/content/market-based-strategies.

Met Office Hadley Centre for Climate Science and Services. n.d. Accessed July 25, 2024. https://wwwpre.metoffice.gov.uk/weather/climate/met-office-hadley-centre/index.

Nixon, R. 1970. "Special Message to the Congress About Reorganization Plans to Establish the Environmental Protection Agency and the National Oceanic and Atmospheric Administration." The American Presidency Project.

Pouffary, S., A. Antonini, F. Gagnon-Lebrun, Y. Touchette, A. Bisiaux, K. Djemouai, and A. Michaelowaet. 2017. "United Nations Framework: Convention on Climate Change: Twenty-third Conference of the Parties (COP23): 6 to 17 November 2017, Bonn, Germany." https://doi.org/10.5167/uzh-143598.

Severinghaus, J. P., T. Sowers, E. J. Brook, R. B. Alley, and M. L. Bender. 1998. "Timing of Abrupt Climate Change at the End of the Younger Dryas Interval from Thermally Fractionated Gases in Polar Ice." *Nature* 391, no. 6663. Accessed October 15, 2024. https://pure.psu.edu/en/publications/timing-of-abrupt-climate-change-at-the-end-of-the-younger-dryas-i.

"Trenberth Talks about Energy Flows and Global Warming." n.d. Skeptical Science. Accessed July 25, 2024. https://skepticalscience.com/print.php?r=122.

Trenberth, K. E. 2013. "The planet is warming, but the warmth just isn't being manifested at the surface." Interview by *The Conversation*, August 14, 2013.

"Yale Experts Explain Climate Anxiety." 2023. Yale Sustainability, March 13, 2023. Accessed October 15, 2024. https://sustainability.yale.edu/explainers/yale-experts-explain-climate-anxiety.

# 5

# Data and Documents

## Data

### Global Land–Ocean Temperature Index

NASA's Global Climate Change, Vital Signs of the Planet website shows the change in global surface temperature compared to the long-term average from 1951 to 1980, as shown in Figure 5.1. The long-term average is marked as 0 and any anomaly, above or below average temperature, is graphed as a departure from the average. Global temperatures began rising around 1910 and have continued since then. The year 2023 was the hottest year on record since recordkeeping began in 1880. The European Union's Copernicus Climate Change Service reported that the eight warmest years on record have occurred since 2014. NASA's analyses generally match independent analyses prepared by the National Oceanic and Atmospheric Administration (NOAA) and other institutions around the world.

Each of the past four decades has been warmer than one that preceded it. Our planet is now 1.2 °C (2.1 °F) hotter than it was since 1850, when emissions of planet-warming carbon dioxide from the burning of fossil fuels became widespread. Berkeley Earth, an independent organization that analyzes environmental data ranked the year 2022 as fifth warmest "We are continuing the long-term warming trend of the planet," said Zeke Hausfather, a researcher at Berkeley Earth. "If you draw a straight line through temperatures since 1970,

---

Water crashes over a street during Hurricane Harvey, one of the costliest US hurricanes on record. Billion-dollar disaster events in the United States are dominated by tropical cyclones, which also result in the highest number of deaths. Droughts rank second. (Eric Overton/Dreamstime.com)

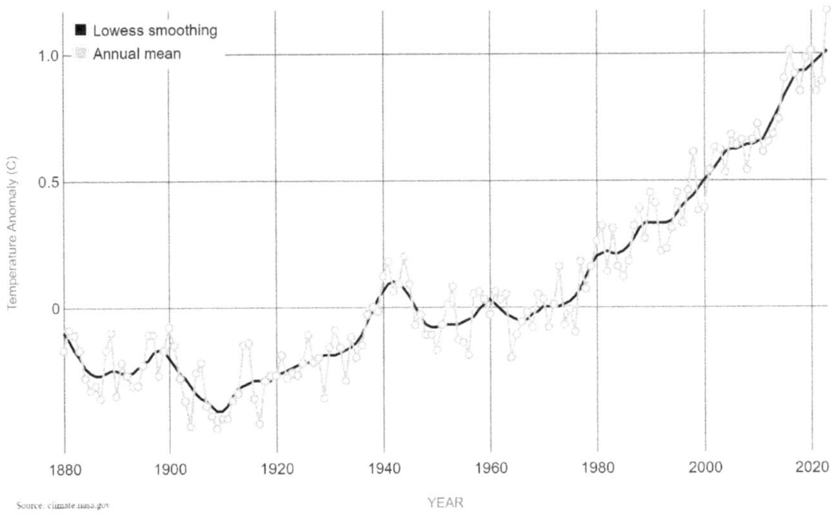


**Figure 5.1** Global Land-Ocean Temperature Index.
**Source:** "Global Land-Ocean Temperature Index." 2023. NASA's Goddard Institute for Space Studies (GISS), July 26, 2023. Accessed November 4, 2024. https://climate.nasa.gov/vital-signs/global-temperature.

2022 lands almost exactly on where you'd expect temperatures to be." Copernicus scientists calculated that Europe had its hottest summer ever in 2022, with several heat waves rolling across the continent that set temperature records in many cities. Several research papers have shown that heat waves in Europe are increasing in frequency and intensity at a faster rate than almost anywhere else. The combination of global warming as well as shifts in atmospheric and oceanic circulation has worsened the heatwaves.

## Estimates of the Earth's Annual and Global Mean Energy Balance

Figure 5.2 depicts a visualization of Earth's energy budget or Earth's radiation budget. Earth, as humans have experienced it, has maintained a balance between the overall amount of incoming and outgoing energy from the sun at the top of the atmosphere. Some of that energy is emitted back to space. The Earth's energy budget stays in balance if incoming energy and outgoing energy are equal. This energy also translates into the Earth's temperatures, because in order for the Earth's temperatures to be stable over long periods of time, the energy budget has to be in balance. In general, if incoming energy is more than outgoing energy, Earth will warm. If outgoing energy is greater than incoming energy,

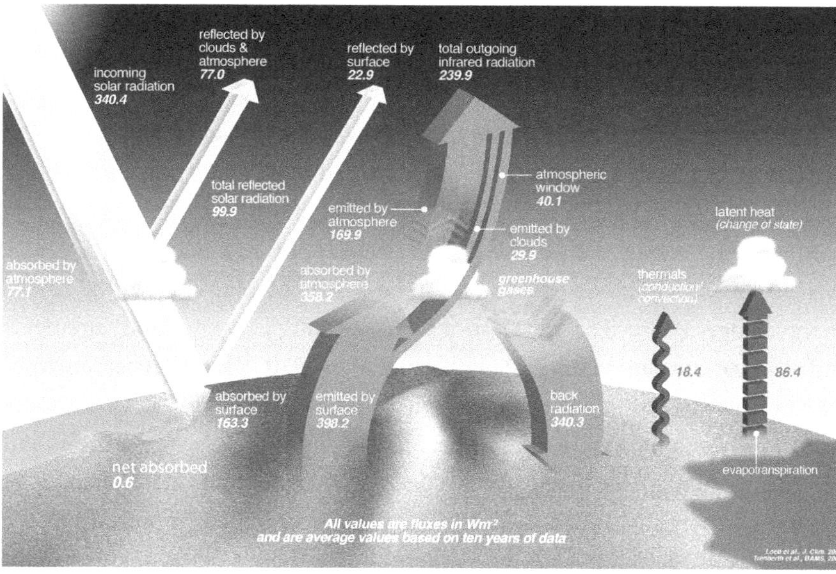


**Figure 5.2** Estimates of the Earth's Annual and Global Mean Energy Balance.
**Source:** "Education Overview." 2022. NASA/CERES, April 8, 2022. Accessed November 5, 2024. https://ceres.larc.nasa.gov/news/education-overview/#additional-energy-budget-resources.

Earth will cool. The temperatures on Earth, especially during the past 11,700 years, or a time geologically referred to as the Holocene, have been pretty stable. Although due to the overabundance of greenhouse gases, we have a surplus of energy being retained.

Incoming solar radiation is measured in Watts per square meter ($W/m^2$). You can see from the large downward-pointing arrow on the left of the figure that the Earth receives about 340 $W/m^2$ per day. About 99.9 ($W/m^2$) of the incoming solar radiation is reflected back out to space, either by clouds and the atmosphere (77.0 $W/m^2$), or by the Earth's surface (22.9 $W/m^2$). When the incoming solar radiation is reflected back to space we call this Earth's albedo or its reflectivity. The word albedo actually comes from the Latin albedo or "whiteness." In science it is used to talk about the fraction of sunlight that is diffusely reflected by a surface. It is measured on a scale from 0 (corresponding to a black body that absorbs all incident radiation and does not reflect) to 1 (corresponding to a body that reflects all incident radiation). A surface with a low albedo does not reflect well and a surface with high albedo reflects well. For example, black tops that absorb heat have a low albedo whereas snow covered surfaces reflect back energy very efficiently and have a high albedo. Low albedo surface can be thought of as retaining more heat and high albedo surfaces tend to stay cooler.

A portion of solar radiation is not reflected back and is absorbed by the various surfaces of the Earth (163.3 $W/m^2$). The absorbed radiation is transformed into long wavelength radiation and is emitted back toward space (398 $W/m^2$). A portion of the energy emitted by the surface escapes back out to space (239.9 $W/m^2$). Greenhouse gases will capture some of the outgoing energy. The Earth's energy budget was well balanced prior to 1970. Multiple types of measurements and observations show a warming imbalance after 1970. The rate of heating from human-caused or anthropogenic warming due to the addition of greenhouse gases into the atmosphere by the burning of fossil fuels, deforestation, and cement production has been unprecedented. The main origin of changes in the Earth's energy is from human-induced changes in the composition of the atmosphere through the continues addition of carbon dioxide, methane, and nitrous oxide. From 2005 to 2019 the Earth's energy imbalance (EEI) averaged $0.90 \pm 0.15$ $W/m^2$.

The energy budget also accounts for how energy moves through the climate system. The tropical areas of the Earth (the areas between the Tropic of Cancer and the Tropic of Capricorn) receive more direct energy from the sun. The standard power on the earth's surface perpendicular to the Sun's rays at sea level on a clear day is approximately 1,000 $W/m^2$. The sun's rays are dispersed and less directly angled toward the poles, so the poles receive much less direct sunlight and less energy (from the tropics to the poles, we get an average of 341 ($W/m^2$)). Sometimes scientists call this differential heating, a fancy way to say differences in heating. As the Earth seeks to balance out the amount of energy across the planet, it drives interactions in Earth's climate system. Imbalances in the Earth's Energy Budget are called radiative forcing. According to NASA, radiative forcing causes chemical and physical changes in the Earth system that impact the climate. An example of this imbalance is the current trend in Earth's Energy Budget; the Earth system currently has a positive forcing since it has more energy entering than leaving.

## Extreme Weather Events in the United States, 1980–2024

Tables 5.1 and 5.2, compiled by the National Oceanic and Atmospheric (NOAA) Center for Environmental Information, are stunning comparison of extreme weather events in the United States from 1980 to 2024. Of greatest concern is the massive jump in the number of events per year from the 1980s, 1990s, and 2000s to a more than doubling of those amounts in the 2010s, and a tripling in the past five years. The number of deaths per year has increased by more than ten times since the 1990s. From 1980 to August 8, 2024, 395 confirmed weather/climate disaster events with losses exceeding $1 billion each affected the United States. These events culminated in the deaths of 16,499 people. The economic effects on the areas impacted and areas serviced by them were substantial. The 1980 to 2023 annual average was 8.5 extreme events (CPI-adjusted); the annual average for the years 2019 to 2023 was 20.4 events (CPI-adjusted). CPI refers to the Consumer Price Index. Adjusting for this allows comparison of costs without inflation distorting the comparison. The top three events to cause fatalities have been tropical cyclones, drought, and severe storms.

**Table 5.1** Select Time Period Comparisons of United States Billion-Dollar Drought, Flooding, Freeze, Severe Storm, Tropical Cyclone, Wildfire, and Winter Storm Statistics (CPI-Adjusted).

| Time Period | Billion-dollar Disasters | Events/ Year | Cost | Percent of Total Cost | Cost/ Year | Deaths | Deaths/ Year |
|---|---|---|---|---|---|---|---|
| 1980s (1980–1989) | 33 | 3.3 | $218.6B | 7.9% | $21.9B | 2,994 | 299 |
| 1990s (1990–1999) | 57 | 5.7 | $333.2B | 12.0% | $33.3B | 3,075 | 308 |
| 2000s (2000–2009) | 67 | 6.7 | $619.2B | 22.3% | $61.9B | 3,102 | 310 |
| 2010s (2010–2019) | 131 | 13.1 | $989.7B | 35.7% | $99.0B | 5,227 | 523 |
| Last 5 Years (2019–2023) | 102 | 20.4 | $617.0B | 22.3% | $123.4B | 1,996 | 399 |
| Last 3 Years (2021–2023) | 66 | 22.0 | $441.4B | 15.9% | $147.1B | 1,690 | 563 |
| Last Year (2023) | 28 | 28.0 | $94.9B | 3.4% | $94.9B | 492 | 492 |
| **All Years (1980–2024)*** | **395** | **8.8** | **$2,772.1B** | **100.0%** | **$61.6B** | **16,499** | **367** |

*Statistics valid as of August 8, 2024.

**Source:** NOAA National Centers for Environmental Information (NCEI). 2024. "U.S. Billion-Dollar Weather and Climate Disasters." Accessed August 9, 2024. https://www.ncei.noaa.gov/access/billions/summary-stats

**Table 5.2** Billion-dollar Events to Affect the United States from 1980 to 2024* (CPI-Adjusted)

| Disaster Type | Events | Events/Year | Percent Frequency | Total Costs | Percent of Total Costs | Cost/Event | Cost/Year | Deaths | Deaths/Year |
|---|---|---|---|---|---|---|---|---|---|
| Drought | 31 | 0.7 | 7.8% | $361.0B | 13.0% | $11.6B | $8.0B | 4,522 | 100 |
| Flooding | 44 | 1.0 | 11.1% | $200.7B | 7.2% | $4.6B | $4.5B | 738 | 16 |
| Freeze | 9 | 0.2 | 2.3% | $37.3B | 1.3% | $4.1B | $0.8B | 162 | 4 |
| Severe Storm | 201 | 4.5 | 50.9% | $503.6B | 18.2% | $2.5B | $11.2B | 2,143 | 48 |
| Tropical Cyclone | 63 | 1.4 | 15.9% | $1,418.2B | 51.2% | $22.5B | $31.5B | 6,934 | 154 |
| Wildfire | 23 | 0.5 | 5.8% | $147.1B | 5.3% | $6.4B | $3.3B | 537 | 12 |
| Winter Storm | 24 | 0.5 | 6.1% | $104.2B | 3.8% | $4.3B | $2.3B | 1,463 | 33 |
| **All Disasters** | **395** | **8.8** | **100.0%** | **$2,772.1B** | **100.0%** | **$7.0B** | **$61.6B** | **16,499** | **367** |

*Statistics valid as of August 8, 2024.

**Source:** NOAA National Centers for Environmental Information (NCEI). 2024. "U.S. Billion-Dollar Weather and Climate Disasters." Accessed August 9, 2024. https://www.ncei.noaa.gov/access/billions/summary-stats

# Documents

## NOAA Atlantic Hurricanes and Climate Change Summary (2023)

*In May of 2023, NOAA addressed three pivotal questions about Atlantic hurricane activity and climate change. The administration's summary first addresses the trend in the number of hurricanes in recent decades. They clarify that even though some measures show a significant increase in hurricane activity since 1980, it is more difficult to determine a trend from the beginning of the twentieth century due to limitations in the earlier data. One of the issues is the lack of consensus in the influence of decreasing amounts of aerosols in the atmosphere since the 1970s on hurricane activity. Records since the 1950s do show clear increase in the number of tropical cyclones that stall off the US coast as well as a clear increase in the rainfall totals since the 1950s. No significant trend has been observed in the number of landfalling tropical storms since the 1900s. NOAA goes on to address two other pressing questions—Has human-caused climate change had any detectable influence on hurricanes and their impacts? And what are the projections for tropical cyclone activity in the Atlantic going forward with continued global warming? NOAA's summary is presented here.*

In terms of important environmental factors related to hurricanes, observed tropical Atlantic and Gulf of Mexico sea surface temperatures show pronounced warming since 1900 as well as multidecadal variability. Since 1950, tropical Atlantic vertical wind shear, sea surface temperatures, and inferred Atlantic Ocean Meridional Overturning Circulation all show pronounced multidecadal variations that are well-correlated to Atlantic major hurricane counts. Economic damage in the US from hurricanes has increased remarkably over the past century, as has the population and value of built infrastructure in hurricane-prone regions. Prehistoric geologic proxy records of hurricane activity covering thousands of years in some locations offer a complementary source of information on the potential for centennial-scale Atlantic hurricane variability originating from natural causes.

### *Attribution of Observed Atlantic Hurricane Changes*

For global mean temperature, the Intergovernmental Panel on Climate Change Sixth Assessment Report made highly confident attribution statements linking global warming to anthropogenic increases in greenhouse gases. The observed global warming trend since the late 1800s stands out from multidecadal variations in the record. This warming trend is consistent with climate model simulations that include anthropogenic increases in greenhouse gases, changes in anthropogenic aerosol forcing, and natural volcanic emissions. In contrast, for Atlantic hurricane activity, the attribution of May 2023 observed changes to increasing greenhouse gases is not yet assessed as highly confident, apart from impacts related to sea level rise.

Observed hurricane data generally either do not show clear centennial-scale trends or do not cover enough years to assess century-scale trends. Pronounced multidecadal variations typically dominate over long-term (centennial-scale) trends over decadal timescales for Atlantic hurricanes. One statistical study concluded that the probability of extreme tropical cyclone rainfall events like that during Hurricane Maria has increased over parts of Puerto Rico since 1956 due to long-term climate change. However, recent observed increases (e.g., since 1980) in Atlantic hurricane activity are generally not representative of longer-term (70-year to centennial-scale) trends, in cases where such longer-term records are available. Rather, trends since 1980 appear to represent the latest upswing in a series of multidecadal variations, leaving open the question of what caused the multidecadal swings.

Two leading candidates for causing the pronounced multidecadal variability in Atlantic basin-wide hurricane activity and related metrics since about 1950 are temporal changes in anthropogenic aerosols and internal climate variability associated with Atlantic Ocean circulation changes. A recent NOAA modeling study demonstrated how the observed increase in Atlantic tropical storm frequency since 1980 can be partly reproduced by the simulated response to reduced anthropogenic aerosol forcing during that time, resulting from more stringent air pollution measures over North America and Europe. Recent NOAA studies have shown that multidecadal variations of tropical North Atlantic vertical wind shear and associated major hurricane counts are well-correlated to a subsurface ocean "fingerprint" of multidecadal variations in the Atlantic Ocean Meridional Overturning Circulation, and that these inferred multidecadal ocean circulation variations are more likely due to internal variability than changes in aerosol forcing. A further linkage has been proposed between multidecadal swings in major hurricane counts and variations in atmospheric dust loading—from natural or anthropogenic sources. Greenhouse gas-induced climate warming may also have contributed to the post-1980 Atlantic hurricane changes, including the observed increase in rapid intensification, which one modeling study links in part to greenhouse gas-dominated anthropogenic forcing. However, confident quantitative attribution of observed multidecadal Atlantic hurricane variability to the above mechanisms remains an unsettled topic of research. Particularly active areas of research include the uncertain modeled response of the Atlantic basin to anthropogenic aerosol forcing and the uncertain magnitude of internal multidecadal variability of the Atlantic Ocean Meridional Overturning Circulation, which is apparently underestimated in many climate models.

Despite difficulties in attributing past multidecadal Atlantic hurricane variations to aerosols, internal natural variability, or other factors, the implications of such attributions are substantial. For example, if the multidecadal Atlantic major hurricane drought in the 1970s and 1980s were primarily aerosol-induced, then one might expect continued elevated Atlantic major hurricane counts in coming decades (in combination with any greenhouse gas-induced trend) since we don't anticipate a resumption of high levels of anthropogenic aerosol emissions from North America or Europe. However, if the Atlantic major hurricane drought were primarily a natural fluctuation due to ocean circulation changes, then another such period of reduced activity might be expected to recur in the coming few decades.

Concerning event attribution, model simulations of current and preindustrial conditions and observed records of extreme rainfall in general have been used to attribute extreme rainfall amounts in several recent tropical cyclone events partly to anthropogenic forcing such as increased greenhouse gases.

There is disagreement among the few published analyses on whether US hurricane damage, when normalized by temporal changes in exposed wealth, has increased since 1900. However, these studies agree that increases in population and wealth, including the value of built infrastructure in hurricane-prone regions, are dominant over hurricane changes in explaining the increase over the past century in annual economic damage from US landfalling hurricanes.

### *Future Atlantic Hurricanes and Global Warming*

A detectable greenhouse gas-induced change in historical observed Atlantic hurricane behavior would raise confidence in projected twenty-first century changes under various greenhouse gas-dominated global warming scenarios. However, since a highly confident attribution has not yet been established for Atlantic hurricanes, future projections rely mostly on climate models alone. This contrasts with the case for global and regional temperatures and sea level rise-related impacts, where projections are relatively much more confident. Based on a survey of existing studies, with regards to future North Atlantic, Caribbean Sea, and Gulf of Mexico tropical storm and hurricane activity, a 2 °C (4 °F) global warming scenario would be expected to lead to the following:

- Storm inundation levels during hurricane surge events will increase due to sea level rise, anticipated to rise by about 2 to 3 ft (0.4 to 0.8 meters) by 2100. This sea level rise will contribute toward significantly more coastal destruction and increased economic damage.
- Rainfall rates within tropical storms and hurricanes are projected to increase by about 15%.
- Numbers of Atlantic hurricanes reaching Category 4 or 5 intensity are projected to increase about 10% but with large uncertainty and with some studies projecting a decrease.
- Total numbers of Atlantic tropical storms and hurricanes combined are projected to decrease by 15%, but with large uncertainty; a minority of studies project an increase.
- Strongest winds of tropical storms and hurricanes are projected to increase about 3%.
- Other aspects of Atlantic hurricanes—such as named storm formation location, tracks, and size—may also change, but there is little consensus in available projections.

**Source:** National Oceanic and Atmospheric Administration (NOAA). 2023. "Atlantic Hurricanes and Climate Change." State of the Science Fact Sheet, May 2023.

## "The Science of Extreme Event Attribution: How Climate Change is Fueling Severe Weather Events," US Senate Committee Meeting on the Environment and Public Works (2023)

*On November 1, 2023, the US Senate's Committee on Environment and Public Works met in Washington, DC. The Senators Thomas R. Carper (Delaware), Shelley Moore Capito (West Virginia), Lawrence Berkeley National Laboratory Scientist of applied mathematics and computational research division, Michael F. Wehner, Chief resilience officer and deputy director of the Resilient Environment Department of Broward, County Florida, and Paul M. Dabbar, the former Under Secretary for Science at the US Department of Energy and research scholar at*

*Columbia University met at the first-ever Congressional hearing on the topic of extreme event attribution. The following excerpts from the meeting are pertinent to the discussions in this book. Specifically, senators ask climate and energy experts about extreme events in their state, the event attribution studies concerning those events, and the future of US energy, infrastructure, mitigation and adaptation.*

*Senator Carper opens the meeting with the following statement:*

First, what is extreme event attribution anyway? Besides being quite a mouthful, extreme event attribution looks at how a specific extreme weather event, such as a particular heat wave or flood, was made worse by climate change.

Second, why is this kind of climate science important, real important? To answer that question, we need to first acknowledge the fact that human-caused climate change is increasing the frequency of extreme weather events. Earlier this year, the American Meteorological Society issued a report by many of our Nation's leading climate scientists and meteorologists. That report detailed, as I am sure our witnesses know, how climate change has driven unprecedented heat waves, floods, and droughts this year and in recent years. We know that continues to be the case.

When Phoenix, Arizona experiences an unprecedented 31 days of temperatures at or above 110 degrees Fahrenheit as they did this summer, or when historic rainfall leads to severe flooding in places like Florida and Vermont as it did this year, many of us find ourselves asking, is climate change to blame for this? The honest answer is yes, yes, it is.

It is true that we have always had heat waves, at least for as long as I have been around, 76 years. It is also true that climate change is making them more intense. So, the better question to ask ourselves is, how much worse did climate change make this heat wave or how much worse did it make that flood? That is the kind of question that extreme event attribution scientists, including at least one of our witnesses, Dr. Wehner, work to answer. These questions matter because the human harms and costs of climate change are massive, and sadly, they are growing.

As many of our colleagues know, I represent the lowest lying State in our Nation, Delaware. In Delaware, we are already losing our dunes, we are losing our wetlands to sea level rise and nor'easters. As we work to repair our homes, businesses, and infrastructure and replenish our beaches damaged by these events, we are already paying for the costs of climate change.

But this hearing is not just about Delaware. It is a hearing about 49 other States and a planet that we all share with people around the world. We are grappling with the costs of climate change.

Today, we are going to focus on how it is fueling extreme weather and, maybe even more important, what we can do about it. Now, in one sense, climate change is already affecting every aspect of our weather. Overall, NASA tells us that global temperatures have increased a little more than 1 degree Celsius, that is about 2 degrees Fahrenheit, since 1880. That is due mostly to human-caused climate change.

Because our planet is warmer, on average, the hot days are becoming hotter. And, because a warmer atmosphere holds more water vapor, rainstorms are growing more intense.

Warmer oceans are also producing stronger hurricanes. Last week, we saw Hurricane Otis wreak havoc, wreak devastation in Mexico when it rapidly intensified from a Category 1 to a Category 5 in less than 10 hours. Think about that. Less than 10 hours. I couldn't believe it, but it is true. Let me close and turn it over to Senator Capito, but first let me say, you and I don't experience planetary averages. We live through and we clean up from and we pay for specific weather disasters. Extreme event attribution science helps us to explain those events.

So, recognizing that we as a planet must prepare for more frequent and intense weather in the future, as elected leaders we also need to understand what to expect in our States and our communities. How much worse will our heat waves become in the years to come? How likely is it that a community will experience an even larger flood next time?

Extreme event attribution science is helping us answer those questions, as well. The ability to do so will be critical as we plan future infrastructure projects, we work a lot on that kind of stuff here in this committee. We wrote big parts of the Bipartisan Infrastructure Law, this lady right here and I, and we had the privilege of managing it on the Floor, one of the biggest infrastructure bills in the history of the Country.

Making more informed policy decisions will help us protect more Americans from extreme weather and allow us to use taxpayer dollars more effectively.

Let me close by reminding everyone that while climate change is driving extreme weather, we are not helpless, and we are not hopeless. The situation is not hopeless. Working together, we can prevent the worst impacts of global warming by reducing greenhouse gas emissions. That is what we are doing. Thanks to the Bipartisan Infrastructure Law that I just mentioned and the Inflation Reduction

Act, we are beginning to turn this adversity into opportunity. Importantly, we are doing so in a way that invests in American-made clean energy, lowers energy costs, creates good-paying jobs all over the Country, and makes communities all over the Country more resilient. Still, it is important to acknowledge that we have a lot more work to do ahead of us to tackle this challenge. So, as we take those steps to better prepare for extreme weather and manage its impacts, extreme event attribution can be an important tool.

With that, we look forward to hearing from our witnesses today about how communities can better understand and anticipate what is coming and increase their resilience to climate-fueled extreme weather.

Before we do that, we will hear from our Ranking Member, Senator Capito, for her opening statement. Senator Capito, welcome. You are recognized. Thank you.

STATEMENT OF MICHAEL F. WEHNER, PH.D., SENIOR SCIENTIST, APPLIED MATHEMATICS AND COMPUTATIONAL RESEARCH DIVISION, LAWRENCE BERKLEY NATIONAL LABORATORY

Mr. Wehner. Thank you, Chairman Carper, Ranking Member Capito, and distinguished members of the committee. Good morning, and thank you for the invitation to testify at this important hearing on the science of extreme weather event attribution.

As you said, my research focuses on the behavior of extreme weather events and the changing climate. I must say that my remarks are my own, and not intended to represent positions of the Lawrence Berkley National Laboratory, the University of California, or the United States Department of Energy.

According to the US National Climate Assessments Annual Report to the Intergovernmental Panel on Climate Change, it is unequivocal that humans have heated the Earth's climate. The best estimate is that human activities, principally the use of oil, coal, and gas are responsible for all of the observed global warming since 1900.

Our understanding about the effects of this human-caused global warming on specific, individual weather events has advanced considerably in the past two decades. For many types of weather events, scientists can identify and quantify the ways that human interference in the climate system has influenced extreme weather.

Obviously, as you said, we have always experienced extreme weather: heatwaves, droughts, extreme storms, and the like. Extreme weather attribution science attempts to quantify the influence of climate change on these specific individual events by answering two related questions. First, has global warming affected the severity of an event of a particular frequency, say, once in 100 years? And second, given the observed intensity of an event, has global warming affected how rare it is? I detailed this a little bit more in my written testimony.

Because we have only one planet Earth, to answer these questions, scientists must use both climate and statistical models to compare representations of weather events in the actual "world that was" to a "world that might have been" without climate change. Confidence is increased when multiple independent research teams use different approaches and arrive at similar conclusions consistent with observed trends.

Confidence in quantitative attribution statements is very high about the human influence on heatwaves, agricultural drought and certain classes of severe storms, including hurricanes. Indeed, as the earth warms, every heatwave that we now experience is hotter than it would have been without climate change, including those this past summer in the United States and throughout the Northern Hemisphere.

Throughout the lower 48 States, I estimate that any heatwave we now experience is two and a half to five degrees Fahrenheit hotter than it would have been. All Americans face increased health risks from these hotter heatwaves, particularly the very young, the very old, and the very poor in our society.

Extreme storms have been made wetter by climate change as well. For instance, more than 30 different hurricanes have been studied in an attribution context. All analysis reveals significant human fingerprint on the total rainfall amounts.

Recent advances in attribution science have gone beyond studying the human influence on the meteorology of extreme weather events to include the impacts of these events on real people. Take Hurricane Harvey and the record flooding it caused in the greater Houston area, for example. Global warming made the surface waters of the Gulf of Mexico about two degrees Fahrenheit warmer, increasing the rainfall during Hurricane Harvey by about 20 percent. This increased the area flooded by about 14 percent, importantly leading to a 32 percent increase in the number of flooded homes in Harris County.

I estimate that global warming is then responsible for about a third of the $150 billion in damages estimated by NOAA during Hurricane Harvey. These damages were not equally distributed across socioeconomic groups. Half of these flooded homes were in low-income, Hispanic neighborhoods. As about a third of Harris County is characterized as low-income Hispanic, this disproportionate impact represents and environmental injustice, in my opinion. We are finding similar injustice to the most vulnerable of our society in an analysis of flooding in New York, New Jersey, Pennsylvania from the remnants of Hurricane Ida.

This recent extension of attribution science from weather to impacts could be informative in negotiation for the Loss and Damages Fund to aid nations particularly vulnerable to climate change. This fund was established but not financed at last year's meeting of the Conference of Parties, the COP27, part of the United Nations framework convention on climate change, and will certainly be one of the topics discussed later this month at the COP28 in Dubai.

The human influence on extreme weather and its impact on people is quite clear. I am often asked, why do we do these attribution studies? I have three answers to this question. First, the public demand for information about how climate change affects them personally is very high. People want to know.

Second, increasing the number and variety of individual extreme events studied increases our understanding of the extent of the human influence on them. Third, and perhaps most importantly, our quantitative understanding can aid decision makers, increasing the resilience of our society to a future hotter world.

Extreme weather event attribution has shown us that dangerous climate change is already happening. How much more dangerous we let it become is up to us.

Thank you, and I welcome your questions.

EXCERPT FROM THE STATEMENT OF JENNIFER JURADO, PH.D., CHIEF RESILIENCE OFFICER AND DEPUTY DIRECTOR, RESILIENT ENVIRONMENT DEPARTMENT, BROWARD COUNTY, FLORIDA

Ms. Jurado. Good morning, Chairman Carper, Ranking Member Capito. Thank you for your leadership in convening this hearing today.

As you shared, I am the Chief Resilience Officer for Broward County, the 17th largest county in the US and the second most populous in the State of Florida, with nearly 2 million residents. While we are leaders in many ways in adapting to a changing climate, we remain at the forefront of significant impacts.

South Florida is no stranger to extreme weather. To manage nearly 60 inches of rainfall we receive annually, we have made extensive investments in drainage, flood control, stormwater management systems. Even so, as we grapple with the impacts of climate change, the limitations of these investments are evident, and record-breaking events are becoming more damaging.

Most recently, on April 12th, 2023, a thunderstorm delivered an unprecedented 26 inches of rainfall in 12 hours, impacting much of our community. A 30-minute commute became a 3-hour navigation of flood waters. Flooded vehicles were abandoned en masse. The Fort Lauderdale-Hollywood International Airport shut down and remained closed for 40 hours. Fuel distribution from Port Everglades was disrupted, affecting 12 counties and 5 international airports. Water levels reached 2 to 3 feet in several older neighborhoods, and the City of Fort Lauderdale remains without a city hall. Yet this event occurred outside the hurricane season, a one in a 1,000 year rainfall event.

Less than three years earlier, Tropical Storm Eta delivered 22 inches of rainfall in three days. The six-week rainfall total was four times the historic average in inland areas, some of which remained flooded for two weeks.

In 2017, an 18-inch rainfall event closed Sawgrass Mills, the region's largest shopping center, for three days. The economic loss was $30 million. These three extreme events all in the last six years account for the highest annual rainfall totals in the last 30, exceeding 88 inches in 2020.

Of course, extreme conditions are not limited to flooding. This year, Broward County experienced 37 days with a heat index over 105 degrees Fahrenheit compared to an average of five days per year historically. We know that extreme heat disproportionately impacts the under-represented residents, outdoor workers, health-compromised individuals, our youth and elders, affecting finances, earnings and physical health.

Climate change is one of the most pressing issues facing our region. Continued leadership and support for transitions to clean energy, alongside aggressive adaptation actions, is critical to our economic and community vitality. We look forward to continued collaborations with our Federal agency partners, and thank you again for the opportunity to speak today.

STATEMENT OF THE HONORABLE PAUL DABBAR, FORMER UNDER SECRETARY FOR SCIENCE, US DEPARTMENT OF ENERGY, SENIOR RESEARCH SCHOLAR, COLUMBIA UNIVERSITY

Mr. Dabbar. Chair Carper, Ranking Member Capito, I am honored to be before the Senate again, for the first time before this committee, to discuss climate impact, energy innovation and policy.

Over my career, I have been engaged in all the various aspects of energy, including liberating neutrons at a reactor, and addressing solutions for the environment, in particular as Under Secretary for Science at the Department of

Energy. Other than Senator Kelly, I have likely traveled to more remote locations to support collecting climate data. I am one of the few people who have been to both the geographic North and South Poles, both in government service, and in part both missions were around climate data gathering.

As Under Secretary, we supported gathering data and computer simulation of the climate, including the work of Dr. Wehner, at many of the national labs I ran including Lawrence Berkeley, as well as at NETL and others in support research at universities in everyone's State here at the committee.

The world has been successful at reducing different types of emissions, due to innovation. And I believe the right strategy for the world today is to continue discovery, innovation and deployment of new options. America has been the dominant country at investing in discovery, and the lead in deployment of new energy technologies, and America is the global leader of the pipeline of new prospects.

I am quite positive about those prospects due to the US being the world's leader in public and private energy R&D, including at the world's leading universities and national labs, the dominant winner of Nobel prizes in the physical sciences, and the global leader in venture capital and start-ups.

I reach this positive conclusion based on a lot of data. Most of what we are deploying today was invented within the last generation, such as commercially effective lithium-ion battery chemistry, wind turbines, innovative drilling technologies, and PV solar. As Under Secretary I enjoyed the Nobel Prize reception in the Senate vestibule for us winning the Nobel for lithium-ion chemistry.

One final innovation take-away that we should consider is that regulators or elected officials mandating winning energy technologies is poor technology policy. The innovation ecosystem has a track record of inventing new options that were not predicted.

For example, President Bush's famous 2001 energy assessment had a long list of predictions of winning energy technologies, and most of them were significantly wrong. Unpredicted innovations in drilling, solar and batteries came along and remade the landscape, including for emissions. It is better policy for the EPA and States to allow for technology neutral, innovation-open strategies, and allow for the competition of discovery.

We need balance in energy policy. We need to concurrently care about increasing energy availability, lowering energy costs, lowering emissions, and national security. Due to American innovation and a solid all-of-the-above bipartisan pro-supply policies over the last generation, we were able to deliver on all those concurrently. I have confidence we can do that again this generation.

While we certainly need to understand the drivers of climate change, we should focus on solutions also. America is the global energy superpower. We went from the largest energy importer in the world to the largest energy exporter. We invented for us and the world a portfolio of technologies that are making an impact on emissions. And there is a pipeline of even better ones to come. Policy to allow the ingenuity of Americans to harness and allow deployment of their new ideas, will be the solution.

[The prepared statement of Mr. Dabbar follows:]

Senator Carper. Great, thank you very much.

Let me start with Dr. Wehner. Your colleague to your right, Dr. Jurado, has told us that Fort Lauderdale and surrounding communities in southeast Florida have experienced several devastating torrential rainfall events over the last couple of years, including 26 inches of rainfall over 12 hours last April. My question, Dr. Wehner, is would you consider this torrential rainfall event unusual? Should we expect to see more extreme rainfall events as climate change worsens?

Mr. Wehner. Thank you, Senator. Yes, indeed, we talked briefly prior to the hearing about this storm. What she said to me about this astonishing amount of rainfall, and most of it falling in a very short period of time, is completely consistent with my previous work on extreme storms, showing that the rainfall increases at a rate greater than humidity increases.

Now, humidity increases at a certain rate determined, that we know, very rigorously from first principles. But rainfall seems to be supercharged from climate change.

So, what I would interpret is that this storm has been made more violent and hence more efficient at increasing this available moisture. We would expect yet more of that throughout the Country, not just in Florida, actually throughout the world.

Senator Carper. How about heat waves? Absent climate change, would we be experiencing events such as the Pacific Northwest 2021 heatwave, or this past summer's heatwave in the US?

Mr. Wehner. It is interesting that you would highlight the 2021 heatwave in the Pacific Northwest. I consider this one a teachable moment for climate scientists. If you had asked me before that event when it would be 120 degrees in Canada, I would have said not for a long time.

Senator Carper. Would you have said when hell freezes over?

Mr. Wehner. No, I would have said about 2060.

[Laughter.]

Senator Carper. That is when it is scheduled to freeze over.

Mr. Wehner. So that was a surprise to us. This will be an event that is probably the most stunning extreme weather event of all time. Currently, there are at least 20 different papers on an event that happened over a year ago. I have three myself, three of those 20 are mine.

We are learning a lot. It is kind of scary, because it was scary beforehand, and clearly this event was unprecedented, causing a lot of people to die. How many more died because of climate change is not something I know.

But I would say that it was at least two degrees warmer from climate change, possibly four, not much more than that.

That doesn't sound like a lot. But when you look at the effect of high temperatures on mortality, a small change from 110 degrees to 114 degrees actually has a large effect on the number of people that die. So that is important.

As one of the things we have seen in the data, that mortality from natural disasters is significantly lower in more technologically advanced societies with access to energy and resilient infrastructure. Recent reports by reliability experts show that we are projected to go, we as a Nation, are projected to go backwards on our grid reliability this decade. Many point to some regulations that will lead to early power plant retirements. We also see the rise of the electric economy that is being moved forward rather rapidly.

Everybody has mentioned that the most vulnerable are those that are in the lower economic echelon of our society. That is troublesome, obviously, for me.

How will extreme weather impact energy demand, and how do you think grid reliability will impact our vulnerability in these events?

We have had near-term emission reductions, but we are still and will continue, I think everybody is in agreement here, to experience climate change impacts until globally our greenhouse gases are reduced. But it may continue then. There is stuff we don't know, that we don't know into our future.

So, given that we helped create this problem we have maybe an oversized obligation into addressing it. I think we are. I think we are, we are doing good work in this panel, we are doing good work with this Administration, and hopefully with the next Administration that comes, much will be put around the Country. Thank you.

Okay, we have been joined by Senator Markey. Senator Markey has been working on these things for a long time, and we are delighted that he has joined us, along with Senator Boozman and Senator Padilla. Senator Markey?

Senator Markey. Thank you. And we know it is not a coincidence that disproportionately Black, Brown, and low-income communities are more likely to struggle with extreme weather events fueled by climate change. Dr. Wehner, in your research, do disasters fueled by climate change hit some neighborhoods harder than others? Does that drive increasing inequality resulting from climate injustices?

Mr. Wehner. Yes, it certainly does. But I would caution that every event, every disaster is different, and affected differently. I think the important point is that the very poor are indeed the most vulnerable, because they are the least able to recover from these kinds of events. So that is pretty obvious.

Senator Markey. Yes, it is obvious. And do you agree with the research that says that extreme weather events are up from three times per year in the United States in the 1980s to 20 per year now in the 2020s? Do you agree with that?

Mr. Wehner. I might have written that.

[Laughter]

Senator Markey. Please expand. Because obviously that is a seven-fold increase in the number of extreme weather events in the United States over the last 40 years. We know it hits poor people disproportionately, although it hits all people who are impacted.

Mr. Wehner. Senator Capito made a point that climate change didn't cause these events. I think that is something I completely agree with. We borrow from epidemiology techniques to sort of understand how climate change has affected these events. Some events might be less frequent, blizzards, for instance, in a warmer world, they become rainstorms. But certainly for heatwaves and violent rainstorms, like hurricanes or this one that hit Fort Lauderdale, are more frequent.

Senator Markey. Let me ask another way. NOAA data shows that disasters that cause more than $1 billion in damage rose from three per year in the 1980s to more than 20 a year in the 2020s. Do you agree with that?

Mr. Wehner. Yes, absolutely.

Senator Markey.. Mr. Wehner, does climate attribution science show that fossil fuel pollution is underpinning this rapid increase in devastating, expensive disasters?

Mr. Wehner. I would agree with that.

Senator Markey. Okay. I think that is the important factor, it has gone from three to 20, it is not all attributable to one thing, but from your research, it is the underpinning of the rapid increase, is that correct?

Mr. Wehner. Yes, it is.

Senator Markey. And then we know that poor people are going to be ultimately disproportionately negatively impacted. Thank you so much, sir.

Senator Carper. Senator Padilla, welcome. You are on. Senator Whitehouse, nice to see you. Thanks for joining us.

Senator Padilla. Thank you, Mr. Chairman. I know I am speaking a little bit to the choir here, but for the record, according to the Fourth National Climate Assessment, which is a report from a Congressionally mandated program, climate attribution studies have indicated that throughout the western United States human-induced climate change is substantially reducing winter and spring snow pack, which is then increasing the likelihood of chronic drought.

The report also indicates that under no change to water management practices, several important western US snowpack reservoirs, including in Sierra Nevada, California, will effectively disappear by the end of this century. Dr. Wehner, as a contributor to the assessment, can you walk us through the impact of snowpack loss for California and other western States?

Mr. Wehner. Yes, Senator, I can. In fact, our research was initially performed by Dr. Alan Rhoades at Berkley Lab when he was a graduate student.

In that we are basically, if we continue on a business as usual or a no policy scenario of emissions, by the end of the century the snow pack in the Sierra would be effectively gone, as it would also in the Wasatch Mountains, which is an important source of water for Utah and Salt Lake City.

So, in California and throughout the west, we rely on the snowpack as a temporary reservoir to supply water for both agricultural needs and for our urban requirements. That infrastructure was built over the past century and was assuming that the snow would melt on a schedule that was in the past. That has changed. There are studies that show that the snow melt has begun earlier. As that continues, it will overwhelm our ability to manage water, so we will run out in the latter parts of the year, before it starts snowing again or raining again.

Senator Padilla. Yes, we are just compounding the effect. This last winter was an anomaly, so we have less precipitation earlier, quicker runoff that leads to very difficult scenarios later in the summer and early fall each and every year, with a growing population and a growing economy. So just laying that out for my colleagues. We need to do different going forward, and we need both policy and investments building on what we have done with the Inflation Reduction Act, building on what we have done with the Bipartisan Infrastructure Law.

Now, speaking of precipitation, I do want to also acknowledge the historic summer events that took place this last winter that illustrated our Nation's vulnerability to heavy precipitation and flooding, when we get the little rain that we do get too much too quickly. Atmospheric river rainstorms are responsible for nearly 85 percent of flooding on the West Coast, which threatens our vital infrastructure and vulnerable communities. In California specifically, we had nine successive atmospheric rivers in January, which caused flooding, power outages and mudslides resulting in an estimated $4.6 billion worth of damages.

I lay that out because it is important to get the data to better understand what we are grappling with here, so we can inform how we adapt and how we respond. So, I was actually glad to see the President's supplemental request funding for NOAA's acquisition of two current hurricane hunter aircraft replacements, which is something that both Senator Graham and I, both members of the committee, pushed hard on.

These plans can help provide real-time data from inside atmospheric rivers to increase vital decision-making which is important to protect life and property. Dr. Wehner, in your testimony you mentioned that climate attribution research in heatwaves and hurricanes is more advanced. Can you talk more about the need for research and development around atmospheric river events?

Mr. Wehner. Indeed, the literature on hurricanes is long and detailed. Although we don't know everything, obviously, about hurricanes. Atmospheric rivers weren't even named such until about 20 years ago. So, it is a much younger field of research and there is much more to be done.

In fact, what you mentioned about sending airplanes out into these storms has only just begun. But it is really critical to get that kind of data, so we understand why some of these disturbing findings that we have found about atmospheric rivers, particularly the ones that are associated with the so called bomb cyclones, have such a sensitive response, large response in their precipitation amounts from climate change.

Senator Capito. [Presiding.] Senator Whitehouse. Senator Whitehouse. Thank you.

Dr. Jurado, I am a Senator from Rhode Island, a fellow ocean State. So I think I will focus my questions on you. I gather you are seeing continuing and increasing tidal flooding, sometimes referred to as sunny day flooding that has nothing to do with a rainstorm, but has to do with sea level rise and tidal action?

Ms. Jurado. Yes, Senator, absolutely. I think these events were first observed with increased regularity about 10 years ago. Then we began to see longer fall tidal flooding. Now, the flooding also occurs into the spring months.

So, you can really set your clock by the expansiveness and frequency of these events. They are particularly compounding flood conditions, because when it rains and we are having this high tide flooding, none of our flat land can drain. We actually find tidal flooding contributing to inland flooding miles inside the county, because of the interconnectedness of these systems.

Senator Whitehouse. A head-up to my next question, well done. So if you are nowhere near the coast in Rhode Island, but you are potentially in a riparian zone, near a river that is going by, or a canal that is going by, you could well be flooded with fresh water because of the sea level rise creating hydraulic back pressure on the escape of that fresh water from Florida's land surfaces?

Ms. Jurado. That is exactly the condition. In fact, some of our most western communities, we see tidal signals in the canal network that are 15 miles inland. So that increase in water level that carries through the entire network is not constrained to the coast by any means. But clearly the most severe impacts are happening in the coastal area. That can extend six miles inland.

And is not only affecting what happens with surface water flooding, but we also find a rise in the groundwater table that is extending six miles inland as well. That affects all of the drainage infrastructure associated with site development, roadways, transportation projects as a whole.

Senator Whitehouse. Along with the rise in the groundwater from effectively hydraulic sea level rise pressure, are you also seeing that salt water is intruding through the limestone into what previously were freshwater wells and water sources for homes and businesses?

Ms. Jurado. That too is absolutely the case. We have estimated that we lost about 30 percent of the coastal wells, or will, to saltwater intrusion by about the year 2050, 2060. We partnered with Palm Beach County to develop a very large surface water reservoir as an alternative water supply, holding 35 million gallons of water to help compensate for what is happening with saltwater intrusion. It is very much driving the practices as we look at long-term water supply investments, and impacts on our wastewater collection system as well.

Because that water is driving into those connections. Senator Whitehouse. Is the warming of the ocean offshore, including measured up to literally jacuzzi recommended temperatures, causing an effect on the strength of storms that hit Florida, particularly hurricane level storms that come from the ocean?

Ms. Jurado. I believe the science is well documented that the warming is causing an intensification of these systems that contributes to additional rainfall that indeed, these storms move more slowly, they carry more rain, they dump more water on communities. So those impacts are very real, in addition to the environmental consequences for the reefs, which are suffering under 100-plus temperatures.

Senator Whitehouse. Thanks very much. I have gone over my time. I appreciate the indulgence of my colleagues. I would only add that when I last checked, Citizens Property Insurance had reserves of $4 billion, Idalia so far is over $10 billion, Hurricane Ian was over $110 billion. So there is an enormous gap there.

Senator Carper. [Presiding.] There certainly is. Thank you, Sheldon. Senator Merkley, good to see you. Welcome. Delighted you have joined us.

Senator Merkley. Thank you very much. I want to continue the impact on insurance. I picture middle class families who have a mortgage, and their mortgage generally says, you must maintain homeowner insurance. So the family gets a notice saying, sorry, we are canceling your homeowner insurance. And the family has to go out and find separate homeowner insurance, otherwise the mortgage company will put the homeowner insurance on a very, very expensive policy. But it becomes increasingly hard to find that.

Senator Merkley. We do see in Oregon a lot of families affected by flood insurance issues. But I really want to focus, when you think about Oregon and climate chaos, you might hear people talk about the pine beetle infestation, you might hear them talk about the decreasing snow pack in the Cascades, which has dropped by an average of 240 inches over 90 years, so basically half the snow pack is gone. You might hear them talk about ocean acidification affecting the sea life off our coast. We have one of the richest grounds in the combination of the Pacific upwelling, the California current.

But the thing that really is on people's minds are fires. We have seen such a growth in forest fires. In 2020, we had three simultaneous fires in three major watersheds. We had the Santiam Canyon fire, the Almeda Fire, the South Obenchain, and Echo Mountain, more than three. Essentially, we lost 4,000 homes. I traveled 600 miles up and down the State, never got out of the smoke. We had massive collapse of communities turning up at the fairgrounds in refugee status. Seeing the burned remains of towns, some 12 different towns were either obliterated or deeply damaged by those fires.

I think maybe this is more for you, Dr. Wehner. People used to talk about 100-year events, they used to talk about 1,000-year events. I think you framed in your testimony that things that occurred every 1,500 years have become once in a century events. Things that are very rare are happening with increasing frequency.

You didn't mention in your testimony wildfires as a category of extreme weather which can be attributed with reasonable certainty. What can climate attribution science tell us about individual wildfires?

Mr. Wehner. It can tell us quite a bit, Senator. I was the chapter lead author on chapter eight of the Fourth National Climate Assessment, Volume One, and fires was part of that chapter. That was back in 2017, and we saw a clear trend, particularly in Washington, Oregon, and California, of large wildfires, an increasing trend. That of course is much worse since 2017, there have been a lot of major fires in all those States.

In anticipation of your question, I did do a little more research last night, trying to find specific things to Oregon. I don't have one. I have some more general statements, though. The leaders in this actually turned out to be Environment Canada and Victoria. They of course studied Canadian fires. But that is not all that different than Oregon.

They found that the fire weather metrics that led to that big fire in 2018 in British Columbia were made two to four times more likely, and more importantly, that the area of the fire was increased by a factor of seven to eleven. So it became a record kind of fire.

Senator Merkley. The fire was seven to eleven times more destructive?

Mr. Wehner. The area burned was seven times larger, at least.

Another study said that nearly all the observed increase of the burned area in California over the past half century is attributable to anthropogenic climate change. So I am sure this is true in Oregon as well. Fire is very difficult, and perhaps one of the more challenging areas for event attribution, because there are, as all these kinds of extreme events, multiple causes. Wind is an instigator, and one of the things that my research and others have found is that there is little trend in extreme wind. So this increase in fire weather conditions is almost entirely due to higher temperatures drawing out the fuel of the forest, so that the fire seasons are longer. This is also reflected in the budgets of firefighters, wildfire fighters. They have to pay more because they are out there longer.

Senator Merkley. We used to, if you asked an Oregonian 20 years ago, when to come to Oregon, they would be like, oh, well, July, August, into early September are our golden time. Now if you ask Oregonians, they will kind of pause and go, you know, we have had a lot of smoke in July and August and September. It is just a huge change. We have furniture stores that can't sell their inventory because of smoke damage, we have wine grapes that have been contaminated by smoke damage, we have people in the hospital with asthma aggravated by smoke, we have communities trying to set up clean air sectors.

In other words, it is so palpable, the change within a few decades is so real. If you are trying to hike the Pacific Crest Trail, there were hikers that used to come through in August, you cannot come through. There are so many fires on the Pacific Crest Trail or near it, threatening it. My wife and I have had to change our plans a number of times for those hikes. You don't want to be in the middle of the wilderness with no cell phone connection or knowledge of fires erupting and be on a very remote trail. It is so many dimensions of our life in Oregon that have been affected. The Ashland Shakespeare Festival, huge challenge with cancelations due to air smoke quality.

So, it is very important for us to be able to capture this. We have seen the fire season grow longer. That one set of fires on Labor Day, 2020 burned over a million acres, and 4,000 homes. A dozen communities were affected. It sent a very powerful messages to people across the Country. They remember Paradise, California.

Well, the devastation was greater in Oregon, but it was kind of like, oh, well, now it is not new, devastating fires that burn a community to the ground, and you see cast iron bathtubs and steel staircases still standing in beds of ashes and molybdenum wheels, where they melted and ran across the ground, creating strange sculptures. It is an architecture that we have just never witnessed before, or a scene we have never witnessed.

So I think understanding this continued impact is incredibly important. Thank you all for bringing your insights to bear on the evolution and dynamics in many dimensions, certainly floods, droughts, and fires. Thank you.

Senator Carper. Thanks for those very thoughtful questions.

*The hearing then goes on a second round particularly discussing how attribution science can inform the design and engineering of infrastructure projects. The process finds specific ways for using this information to help the public better understand how climate change impacts their daily lives. The group then discusses how the Inflation Reduction Act programs boost resiliency.*

*Senator Carper adds at the end of the meeting:*

I think we know where we are going if we don't work together, pull together. We use science to guide us. I must say, I wish you could all have been here, I have been on this committee 22 years, I wanted to be on this committee when I got here 22 years ago. But to have heard the kinds of conversations we had then on climate change compared to what we had in this incredibly constructive conversation. I applaud each of you for taking the time to join us. Everything, as proud as I am of the work we have done in the Inflation Reduction Act and the work we have done on the climate change provisions within the Bipartisan Infrastructure Law, everything I do, everything we do we can do better. It is one thing to write legislation and get it signed into law. The really important piece now is implementation.

Folks around the country think we don't like each other, Democrats and Republicans hate each other, we never want to work together. I have people who say to me, at least every week, Sheldon, when I am getting on the train station and getting on a train here or up in Delaware, people say, can't you guys just work together? Can't you work together? Can't you get something done?

I wish they could have sat through this hearing. They might have been encouraged a little bit, because I think we have a pretty good idea of what is wrong. And we have a lot of good ideas about how to address it. If the American people would hear that, I think they would be encouraged.

There was once a woman who ran for President not long ago, and her campaign slogan was stronger together. Stronger together. And I think that really applies here. I think that we are largely together, and we are stronger because of that. I just love getting things done. I love getting things done, and we are getting some good things done. If we keep this attitude going, we will do better still.

Thank you for what you do with your lives, and thank you for joining us today.

With that, a fellow who does great things with his life, from a small State, but as we know from Delaware, small States do great things, and they produce, at least in his case, in his State's case, great legislators.

**Source:** "The Science of Extreme Event Attribution: How Climate Change Is Fueling Severe Weather Events." 2023. U.S. Senate Committee on Environment and Public Works, November 1, 2023. Accessed November 4, 2024. https://www.epw.senate.gov/public/index.cfm/2023/11/the-science-of-extreme-event-attribution-how-climate-change-is-fueling-severe-weather-events.

6

# Resources

This chapter provides an annotated bibliography to guide readers in finding current information on various topics related to extreme weather and climate change. Many of the books suggested will offer an introduction to the concepts discussed in this book. Other books are narratives by top scientists documenting extreme weather events, attribution science developments, and geoengineering efforts to mitigate or adapt to global warming. While most resources provided are books and journal articles, the chapter also points the reader to websites, videos, news reports, and podcasts that are updated frequently (often weekly). The topics covered by these references range from the science of climate change, examples of extreme weather events linked by attribution science, the ethics of climate change, and the vulnerable populations who suffer most from extreme weather events. Many publications are landmark papers that established the connection between extreme weather and its relation to current anthropogenic global warming. The chapter organizes the resources into sections. The first section of the chapter focuses on resources that will further guide the reader on science topics in climate change and extreme weather, and the articles and websites explain the influence that global warming has had and is having on the Earth's systems. The second set of resources focuses on the studies and resources available for those interested in extreme weather's effects on society and the mitigation and adaptation status of various regions. The last sections of the chapter offer sources for interactive atlases, videos, and podcasts that supplement the readings.

---

A woman waiting for relief after Cyclone Aila devastated the Sundarban Delta in India. The poorest nations are often the most vulnerable to extreme weather events resulting from climate change. Within developed countries, Indigenous peoples, ethnic and racial minorities, migrants, children, people with disabilities, and single parents are more susceptible to greater suffering from extreme weather than the average population. (Samrat35/Dreamstime.com)

## Science of Climate Change and Extreme Weather

### Books

Ahrens, D. C., P. J. Samson, K. A. Reed. 2025. *Extreme Weather and Climate*. Boston: Cengage.

*Extreme Weather and Climate* is a focused study on extreme weather that is understandable to those whose background may not be in the field of meteorology or climate science. It discusses the causes and impacts of extreme weather, looking at meteorology and climate science in the context of significant weather events, such as Hurricane Sandy. It also looks at the probability of climate change influencing the frequency, intensity, or duration of extreme weather events. The publisher compliments the book with exciting photos and illustrations, describing the intensity of weather and its destructive impacts on life.

Ahrens, Donald C., and Robert Henson. 2022. *Meteorology Today: An Introduction to Weather, Climate, and the Environment*. Boston: Cengage.

Donald Ahrens and Robert Henson's book on weather, climate, and Earth science is a perfect reference for those looking for background information on meteorology, climate science, and weather's regional implications. The book aims to help readers synthesize information by leading them to analyze meteorological concepts and issues. It includes complete explanations of weather systems, including the science behind cyclone formation, cloud types, fronts, pressure systems, and everyday components of the weather, such as wind, precipitation, condensation, and the seasons. The book discusses new developments in climate patterns and reviews the findings of the US National Climate Assessment and the Intergovernmental Panel on Climate Change, among others.

Bedford, D., and J. Cook. 2016. *Climate Change: Examining the Facts*. Santa Barbara, CA: ABC-CLIO.

Based on the reality that climate change is one of the most misunderstood issues of the twenty-first century, this book takes on common misinformation with a clear explanation of scientific explanations and clarifications. The book addresses questions at the heart of public debates surrounding human-caused global warming. The authors thoroughly dissect thirty-five specific climate change topics that both scientists and climate change skeptics commonly iterate. The descriptions help the reader understand where the scientific evidence is on each subject and the root of any misunderstandings by those who are not climate experts. Bedford and Cook examine the evidence of anthropogenic climate change, the accuracy of climate forecasting models, whether extreme weather events are increasing due to climate change, and the benefits and drawbacks of several proposals to limit greenhouse gas emissions. Each chapter is based on one of the topics and contains several questions that the public and the media often have on the details of the issue. The questions within the topic are addressed and clarified with scientific facts, figures, and further reading suggestions.

Cronin, T. M. 1994. *Paleoclimates: Understanding Climate Change Past and Present*. New York: Columbia University Press.

This book examines four billion years of the Earth's paleoclimatology (the study of past climate changes). It is a collection of important research looking at the physical, chemical, and biological clues, such as glacial ice, tree rings, sediments, corals, and speleothems, imputed into computer models to decipher ancient climate shifts. The reconstructed climate outputs test hypotheses about the causes of climatic change, such as greenhouse gases, solar variability, Earth's orbital variations, and hydrological, oceanic, and tectonic processes. Cronin's book emphasizes discoveries about past periods of global warming and cooling. It also highlights evidence of atmospheric greenhouse gas concentrations, abrupt climate and sea-level change, natural temperature variability, and other topics directly relevant to the causes and impacts of climate change.

Although this book was published more than twenty years ago, it is still used as a reference by many students who seek to understand the many causes of natural climate change. The book caters to advanced undergraduate and graduate students and researchers. Still, it can also be an excellent reference for those requiring a general background in natural climate variability. It also offers introductory material on the climate system and the methods used in paleoclimatology studies. It may be too in-depth for those with little climate science knowledge, but it is a valuable book to have to reference specific changes in climate throughout Earth's history, their causes, and the evidence from which the theories were established.

Dessler, A. E. 2022. *Introduction to Modern Climate Change*. Cambridge, UK: Cambridge University Press.

Dessler's introductory textbook for science and non-science students dives into the pressing developments in recent extreme weather events and human migration due to drought, wildfire, and heat. It reflects on the science of anthropogenic climate change and the policies needed to avoid even worse climate conditions. The book's first half covers the science of modern climate change, including scientific evidence that the Earth is warming and provides a background in climate physics. The book explains fundamental concepts in climate science like radiative forcing, feedback loops, and the carbon cycle. Introductory physics and algebra explain changes in the Earth system, which offers a basic explanation to those wanting a mathematical background on the mechanics of climate change. The chapters cover the basics of climate vs. weather and the substantial evidence that the climate is indeed changing due to human interactions with the atmosphere. The book's second half addresses the economics and policy options available up to 2022 to address climate change. Dessler looks at the predictions of future climate change, the impacts of the change on humanity, the social cost of carbon, and the fundamentals of mitigation and adaptation to extreme weather events resulting from global warming.

Fares, A., ed. 2021. *Climate Change and Extreme Events*. Prairie View, TX: Elsevier.

*Climate Change and Extreme Events* uses a multidisciplinary approach to discuss the relationship between climate change-related weather extremes and the impacts already witnessed on human lives. The book has four major sections: weather parameters, hydrological responses, mitigation and adaptation, and governance and policies. These topics cover past, present, and future perspectives. Sections give an overview of weather mechanics and how the water cycle responds to the changes brought on by global warming. The book presents current knowledge and a future outlook on air and river temperatures, precipitation, storms and hurricanes, flooding, and ecosystem responses to these extremes. Other sections cover extreme weather events and policymakers' responses to the events. This book provides a multisubject approach to those interested in climate science, weather, environmental issues, and social science related to extreme weather.

Fletcher, C. 2019. *Climate Change: What the Science Tells Us*. Hoboken, NJ: John Wiley and Sons.

Charles Fletcher's text is a complete reference to many topics on climate change, including an entire chapter on the impacts of warming on precipitation patterns, extreme rain, snowfall, air temperatures, extreme heat, drought, storm generation, and wildfire. The book also considers the changes these new patterns have had on transportation, agriculture, water resources, species' geography, and human health. The book introduces climate change fundamentals and essential concepts. The book looks at the regional impacts documented as of 2019. It can document early thought regarding innovations and policies considered necessary in the mid-2000s for reaching Paris Agreement goals. Fletcher discusses peer-reviewed literature on climate change on land and sea, the hydrologic cycle, human societies, weather, and future societal projections. The basics of greenhouse gases, oceanic and atmospheric processes, Pleistocene and Holocene paleoclimate, sea levels, and other fundamental topics provide a deep understanding of critical concepts in the field. His discussion of extreme weather and the economic impacts that will continue to increase in the future is a perfect reference for past and current outcomes.

Hakim, G. J., and J. Patoux. 2022. *Weather, a Concise Introduction*. Cambridge, UK: Cambridge University Press.

Gregory Hakim and Jerome Patoux are part of a world-renowned team from the University of Washington. They have written a fully illustrated text covering the foundations of meteorology clearly and engagingly. The authors cover the physical sciences, including the chemistry and physics of the atmosphere, for students with both a beginning understanding of the science and a more advanced understanding. The authors use a case study of a midlatitude cyclone throughout the book to explain key concepts driving the dynamics of the atmosphere. A complete reference for the processes involved in the water cycle, cloud formation, precipitation, wind production, storm generation, tropical cyclones, weather forecasting, and climate change related to extreme weather.

Kostigen, T. M. 2020. *Hacking Planet Earth: How Geoengineering Can Help Us Reimagine the Future*. New York: Penguin Random House.

Thomas Kostigen, a writer for *USA Today* and the *New York Times*, and *New York Times* bestselling author, offers a glimpse at the geoengineering projects discussed by the sharpest minds at the most prestigious institutions to keep global warming from causing irreversible further harm. The book covers topics to solve warming temperatures, including increasing cloud cover, putting shades in outer space, using deserts to reflect incoming radiation, using smart soils, adding additives to the oceans to increase photosynthesis, and other top ideas.

Kostigen interviews experts, designers, and researchers to explain the ideas proposed in what seems like science fiction but is funded science research.

Ongoma, V., and T. Hossein 2022. *Climate Impacts on Extreme Weather: Current to Future Changes on a Local to Global Scale.* Amsterdam: Elsevier.

This book offers up-to-date coverage of climate research related to extreme weather events. The book offers a comprehensive look at the understanding of the drivers of current changes in the climate and how extreme weather events are generated from current climatic conditions, and how the frequency, intensity, and duration of climate extremes have changed in the past few decades. It provides regional and global case studies, giving context to the application of the science. For example, the book looks at features of the Indian monsoon and rainfall extremes, the unpredictability of rainfall over West Africa, and the intensification of rainfall in the United States due to global warming, to name a few. Primary research and models present both observed and projected scenarios of extreme weather events.

Otto, F. 2019. *Angry Weather: Heat Waves, Floods, Storms, and the New Science of Climate Change.* Vancouver: Greystone Books.

This non-fiction, narrative style book written by Friederike Otto, a physicist, philosopher, climate researcher, and lecturer at the Imperial College in London, pinpoints the recent extreme weather events that the World Weather Attribution has attributed to human-caused climate change. The book introduces attribution science, how it works, and its importance. He emphasized the facts that attribution science provides and the responsibility that heavy emitting countries have to those impacted by extreme weather events, especially those statistically known to be larger, more extensive, and deadlier than they would have been without current global warming.

Ruddiman, W. F. 2014. *Earth's Climate Past and Future.* New York: W.H. Freeman and Company.

Written by William Ruddiman, a marine geologist, paleoclimatologist, and professor at the University of Virginia, this book looks at climate changes on planet Earth from natural forces to human activity. Motivated by advances in the field of climate science, he uses a multidisciplinary approach to address evidence for anthropogenically driven global warming. He explores crucial advances in the understanding of past climates. The book looks at climate change through time scales from fifty million years to historical and future scenarios. It also discusses man's role as early as six thousand years ago as we settled and began farming, changing the ecology of the Earth. Ruddiman acknowledges the evidence showing that every human activity is the primary cause of modern global climate change. This book has an excellent section for those particularly interested in the mechanisms involved in monsoonal circulation and the drivers of changes in the North African summer monsoon. The book also provides exceptional coverage of climate change since 1850, including climatic interactions and feedback.

## Articles and Reports

Buis, A. 2021. "The Climate Connections of Record Fire Year in the U.S. West." NASA, February 22, 2021. Accessed October 16, 2024. https://climate.nasa.gov/ask-nasa-climate/3066/the-climate-connections-of-a-record-fire-year-in-the-us-west.

In NASA's Global Climate—Vital Signs of the Planet, Alan Buis examines the fire events of 2020 and the extreme weather events that bore the "fingerprints of climate change." He especially focuses on the western United States. He discusses the reports by CAL FIRE—California's Department of Forestry and Fire Protection Agency. Graphs and maps show the trends in the annual number of large fires in the western United States from 1984 to 2011. Buis looks at the combined effects of climate change-induced drought and increased heat and discusses how the carbon emissions from wildfires impact the climate. This article offers a unique insight into the connection between fire science and climate change science.

Cho, R. 2021. "Attribution Science: Linking Climate Change to Extreme Weather." Columbia Climate School, Climate, Earth, and Society, October 4, 2021. Accessed October 16, 2024. https://news.climate.columbia.edu/2021/10/04/attribution-science-linking-climate-change-to-extreme-weather.

Renee Cho of the Columbia Climate School, Climate Earth, and Society, looks at attribution science. The science that models how much human-induced or anthropogenic greenhouse gases increases the chances or likelihood of the extreme weather events. Cho discusses the findings from various models applied to the European heatwaves of 2003, for example. She also describes the methods of attribution studies and their use by the World Weather Attribution initiative, an organization of international scientists that use real-time models to analyze extreme weather events and their relation to global warming. Models can also analyze events to understand the influence of global warming.

Colbert, A. 2022. "A Force of Nature: Hurricanes in a Changing Climate." NASA Global Climate Change, Vital Signs of the Planet, June 1, 2022. Accessed October 16, 2024. https://climate.nasa.gov/news/3184/a-force-of-nature-hurricanes-in-a-changing-climate.

Angela Colbert, a science writer, explains, in easy-to-understand language, the effects of climate change on hurricane formation in a warming world. In the article, she discusses the latest information from climate models that project increased precipitation from hurricanes generated in warmer waters and a warmer atmosphere. The added precipitation increases the amount of flood risks to areas in the path of a hurricane. Interestingly, the findings at the time of this paper were that the global frequency of storms may decrease or remain unchanged. However, if a cyclone forms, it becomes more intense quickly and has more precipitation output than past hurricanes.

Davis, K. T., S. Z. Dobrowski, P. E. Higuera, Z. A. Holden, T. T. Veblen, M. T. Rother, S. A. Parks, A. Sala, and M. P. Manetaet. 2019. "Wildfires and Climate Change Push Low-Elevation Forests across a Critical Climate Threshold for Tree Regeneration." *Proceedings of the National Academy of Sciences* 116, no. 13 (March 11): 6193–8. https://doi.org/10.1073/pnas.1815107116.

This article, written by respected and established forest fire ecologists, examines the effects of climate change on low-elevation forests due to climate change and the increase in wildfires. After studying thirty-three wildfire locations and the success of seedling establishment, they conclude that the combination of a warmer climate and unfavorable soil moisture after wildfires prevent seedlings of ponderosa pine and Douglas fir trees from surviving. This may lead to ecosystem shifts in low-elevation ponderosa pine and Douglas fir forests across the western United States.

Emanuel, K. 2024. "Cyclone Jasper's Rains in the Context of Climate Change." *Proceedings of the National Academy of Sciences* 121, no. 15 (April). https://doi.org/10.1073/pnas.2400292121.

This paper by Kerry Emanuel, one of the leading researchers of the effects of climate change on hurricanes, investigates the impact of climate change on tropical cyclone rainfall. In this study he investigated cyclone Jasper which struck northern Queensland on December 2023 causing extensive flooding due to torrential rain. Some stations reported rainfall totals exceeding 1 meter, possibly making Jasper the wettest tropical cyclone in Australian history. To estimate the probability of rainfall events like Jasper's, the study applied an advanced tropical cyclone downscaling technique to nine global climate models. The annual probability of 1 meter of rain from tropical cyclones at Cairns increased from about 0.8 percent at the end of the twentieth century to about 2.3 percent at the end of the twenty-first century—a factor of almost three. The primary reasons for increasing rainfall are stronger cyclones and a moister atmosphere. Cyclone Jasper's extreme rainfall highlights the need to better understand and prepare for such events in a changing climate.

"Extreme Heat in North America, Europe and China in July 2023 Made Much More Likely by Climate Change." 2023. World Weather Attribution, July 25, 2023. Accessed October 16, 2024. https://www.worldweatherattribution.org/extreme-heat-in-north-america-europe-and-china-in-july-2023-made-much-more-likely-by-climate-change.

Authors from the Grantham Institute, Imperial College London, the Royal Netherlands Meteorological Institute, and the Red Cross Red Crescent Climate Centre reported the main findings after the 2023 world heatwaves. They found that the 2023 heatwaves were among the deadliest natural hazards not only in 2023 but year after year leading up to the unprecedented year. They argue that the main impact of heatwaves is not fully known due to a lack of extended records and that global mortality figures are likely underestimated. They also point out that the Intergovernmental Panel on Climate Change (IPCC) is expecting more of these heatwaves, with human-caused climate change, making extreme heatwaves expected approximately once every fifteen years in the US/Mexico region, once every ten years in Southern Europe, and once every five years for China.

They stress that extreme heat events would be extremely rare without human-induced climate change. A heat wave like the one that impacted China in 2023 would have occurred once in two hundred and fifty years. The maximum heat in July 2023 would have been virtually impossible in the US/Mexico region and Southern Europe if humans had not warmed the planet by burning fossil fuels. Similar to previous studies, they found that the heatwaves were 2.5 °C warmer in Southern Europe, 2 °C warmer in North America, and about 1 °C in China with anthropogenic global warming. They conclude that these events will become even more common unless the world rapidly stops burning fossil fuels, with hotter and longer-lasting heat waves. Extreme heatwaves could occur every two to five years in a world that is 2 °C warmer than the preindustrial climate. They discuss the efforts of many regions worldwide in implementing heat action plans and the evidence that those efforts have decreased heat-related mortality. They also review the cities with urban planning for extreme heat and report that they tend to be cooler and reduce the urban heat island effect. The authors call for an urgent need to accelerate heat action plans to combat risks driven by the culmination of climate change, population aging, and urbanization.

Glenn, E., D. Comarazamy, J. E. González, and T. Smith. 2015. "Detection of Recent Regional Sea Surface Temperature Warming in the Caribbean and Surrounding Region." *Geophysical Research Letters* 42, no. 16 (August 22): 6785–92. https://doi.org/10.1002/2015gl065002.

This peer-reviewed article shows a sea surface temperature (SST) warming trend for the Caribbean and surrounding region over 1982 to 2012. A thirty-year climatological analysis recorded annual, monthly, and seasonal trends. Results show that SSTs increased every year in the Caribbean. Their study had a high statistical significance. The study also found that warming is greatest in the Gulf of Mexico and north of South America during the early and late rainy seasons. The authors review regional oceanic warming. Other studies have, of course, linked warmer temperatures to storm formation. This study shows that warm SSTs correlate with temperature and precipitation in the tropical regions between North and South America. The paper also discusses the effects of El Niño on SSTs.

Intergovernmental Panel on Climate Change (IPCC). 2021. "Summary for Policymakers." In *Climate Change 2021: The Physical Science Basis. Contribution of Working Group I to the Sixth Assessment Report of the Intergovernmental Panel on Climate Change*, 3-32. Cambridge, UK: Cambridge University Press. https://doi.org/10.1017/9781009157896.001.

This "Summary for Policymakers" reports the key findings of the Working Group I (WGI) contribution to the Intergovernmental Panel on Climate Change (IPCC) Sixth Assessment Report (AR6) on the physical science basis of climate change. The report focuses on the state of the oceans, regions of the planet, and the interaction between the atmosphere and oceans concerning global warming. The report builds upon the findings from previous reports and adds new evidence from climate science. The report provides an extensive summary of the understanding of the current state of the climate, how it is changing and the role of human contributions to greenhouse gas increases, the level of confidence in the knowledge about possible climate future projections, climate information by regions and sectors, and the possibilities and challenges to keeping temperatures from rising above 1.5 °C.

Intergovernmental Panel on Climate Change (IPCC). 2021. "Weather and Climate Extreme Events in a Changing Climate." In *Climate Change 2021: The Physical Science Basis. Contribution of Working Group I to the Sixth Assessment Report of the Intergovernmental Panel on Climate Change*, 1513-766. Cambridge, UK: Cambridge University Press. https://doi.org/10.1017/9781009157896.013.

The IPCC's Sixth Assessment Report's "Chapter 11: Weather and Climate Extreme Events in a Changing Climate" assesses regional and global scale changes in weather and climate extremes. The chapter also includes attributions to current global warming. Scientists also project future changes to regional weather depending on warming scenarios ranging from an increase of 1.5 °C to 4 °C. The projected changes include temperature extremes, heavy precipitation and rain-instigated floods, river floods, droughts, and tropical and midlatitude storms.

Meadows, R. 2024. "Why California Is Being Deluged by Atmospheric Rivers." *Scientific American*, February 20, 2024. Accessed October 16, 2024. https://www.scientificamerican.com/article/why-california-is-being-deluged-by-atmospheric-rivers.

This paper discusses the phenomenon of atmospheric rivers and their impact on California's weather. As explained in Chapter 1 of this book, atmospheric rivers are long, narrow bands of moisture-laden air that transport large amounts of water vapor from the tropics to higher latitudes (toward the polar regions). In California, these atmospheric rivers play a crucial role in delivering much-needed precipitation, but they can also lead to intense rainfall, flooding, and landslides. The article explores the science behind atmospheric rivers and their connection to California's climate patterns.

Meehl, G. A., C. Tebaldi, G. Walton, D. Easterling, and L. McDaniel. 2009. "Relative Increase of Record High Maximum Temperatures Compared to Record Low Minimum Temperatures in the U.S." *Geophysical Research Letters* 36, no. 23 (December). https://doi.org/10.1029/2009gl040736.

This article helps quantify the amount of record high maximums and low minimums as of the end of September 2009, using the data collected by the National Climatic Data Center website that archives temperatures from weather stations across the United States. The authors discovered that the ratio of high maximums to low minimums was roughly two to one. The authors documented that more record high maximum temperatures were set than record low minima simply because the annual US average surface temperatures had increased since the 1970s. They look at observations over the last 50-plus years and model simulations that allow them to study the effects of future warming. They used a subset of nearly 2,000 stations of the over 5,000 quality-controlled network station observations of daily maximum and minimum temperatures. All station records span the same period, from 1950 to 2006. Their results present hundreds of station data totals over continental United States or eastern and western regions. They also used model data from the NCAR Community Climate System for twentieth century and twenty-first-century climate simulations. They included natural (volcanoes and solar) and anthropogenic (GHGs, ozone and direct effect of sulfate aerosols).

Payne, A. E., M.-E. Demory, L. R. Leung, A. M. Ramos, C. A. Shields, J. J. Rutz, N. Siler, G. Villarini, A. Hall, and F. M. Ralph. 2020. "Responses and Impacts of Atmospheric Rivers to Climate Change." *Nature News*, March 9, 2020. Accessed October 16, 2024. https://www.nature.com/articles/s43017-020-0030-5.

This article looks at critical points in the link between atmospheric rivers and a warming world. It summarizes that atmospheric rivers have the vital function of transferring atmospheric moisture from the tropics, or areas between the Tropic of Cancer and the Tropic of Capricorn, to higher latitudes. Once the atmospheric rivers reach land, they can be beneficial in bringing needed rain or disastrous, causing much harm by producing floods, landslides, and extreme rainfall. The laws of thermodynamics come into play in the link between a warming climate and the increased amount of moisture in these systems.

Rhoades, A. M., A. D. Jones, A. Srivastava, H. Huang, T. A. O'Brien, C. M. Patricola, P. A. Ullrich, M. Wehner, and Y. Zhou. 2020. "The Shifting Scales of Western U.S. Landfalling Atmospheric Rivers Under Climate Change." *Earth & Environmental Systems Modeling* (EESM), September 22, 2020. Accessed October 16, 2024. https://climatemodeling.science.energy.gov/research-highlights/shifting-scales-western-us-landfalling-atmospheric-rivers-under-climate-change.

This paper uses Community Earth System Model simulations validated and projected under climate change to show a shift from atmospheric rivers shifting from "mostly or primarily beneficial" to "mostly or primarily hazardous" with an intensification of winter season precipitation totals. Changes in precipitation totals are due to a significant increase in the prevalence of atmospheric river precipitation (+260 percent) through increases in the most intense category of events and a decrease in the interval between the atmospheric rivers making landfall in the western United States. This study is evidence that anthropogenic climate change has and continues to alter the characteristics of atmospheric rivers, leading to various societally relevant impacts. The state-of-the-art global climate simulation tools allowed these researchers to connect global-to-regional scales to better represent the interactions between atmospheric rivers and mountains when they make landfall. The paper also explains the characteristics of the intensity level of the categories of atmospheric ratings.

Sayedi, S. S., B. W. Abbott, B. Vannière, B. Leys, D. Colombaroli, G. Gil Romera, M. Słowiński, J. C. Aleman, O. Blarquez, A. Feurdean, K. Brown, T. Aakala, T. Alenius, K. Allen, M. Andric, Y. Bergeron, S. Biagioni, R. Bradshaw, L. Bremond, E. Brisset, J. Brooks, S. O. Brugger, T. Brussel, H. Cadd, E. Cagliero, C. Carcaillet, V. Carter, F. X. Catry, A. Champreux, E. Chaste, R. D. Chavardès, M. Chipman, M. Conedera, S. Connor, M. Constantine, C. C. Mustaphi, A. N. Dabengwa, W. Daniels, E. De Boer, E. Dietze, J. Estrany, P. Fernandes, W. Finsinger, S. G. A. Flantua, P. Fox-Hughes, D. M. Gaboriau, E. M. Gayo, M. P. Girardin, J. Glenn, R. Glückler, C. González-Arango, M. Groves, D. S. Hamilton, R. J. Hamilton, S. Hantson, K. A. Hapsari, M. Hardiman, D. Hawthorne, K. Hoffman, J. Inoue, A. T. Karp, P. Krebs, C. Kulkarni, N. Kuosmanen, T. Lacourse, M.-P. Ledru, M. Lestienne, C. Long, J. A. López-Sáez, N. Loughlin, M. Niklasson, J. Madrigal, S. Y. Maezumi, K. Marcisz, M. Mariani, D. McWethy, G. Meyer, C. Molinari, E. Montoya, S. Mooney, C. Morales-Molino, J. Morris, P. Moss, I. Oliveras, J. M. Pereira, G. B. Pezzatti, N. Pickarski, R. Pini, E. Rehn, C. C. Remy, J. Revelles, D. Rius, V. Robin, Y. Ruan, N. Rudaya, J. Russell-Smith, H. Seppä, L. Shumilovskikh, W. T. Sommers, Ç. Tavşanoğlu, C. Umbanhowar,

E. Urquiaga, D. Urrego, R. S. Vachula, T. Wallenius, C. You, and A.-L. Daniau. 2024. "Assessing Changes in Global Fire Regimes—Fire Ecology." SpringerOpen, February 8, 2024. Accessed October 16, 2024. https://fireecology.springeropen.com/articles/10.1186/s42408-023-00237-9.

Fire scientists and paleoecologists from around the world studied how fires have changed globally over the Holocene or the last 11,000 years of Earth's history. They identified the main reasons behind shifts in fire patterns, including how often and where fires occur. The study explored how future fire regimes might affect services provided by ecosystems (like clean water, air quality, etc.). Policies that capture and store carbon (like planting trees) need careful evaluation. Expert opinion and climate models project more fire activity in a warming world. This is of concern as fires release carbon back into the atmosphere. The authors urge that to bridge science and policy, governments should involve a diverse expert group, including scientists, local managers, and traditional native land knowledge holders, in decision-making. This helps manage complex fire-related issues effectively. Understanding fire patterns is crucial for protecting our environment and making informed decisions.

Senande-Rivera, M., D. Insua-Costa, and G. Miguez-Macho. 2022. "Spatial and Temporal Expansion of Global Wildland Fire Activity in Response to Climate Change." *Nature Communications* 13, no. 1: 1208. https://doi.org/10.1038/s41467-022-28835-2.

This article looks at several areas of the world affected by wildland fire activity due to climate change. The authors report that global warming is expected to alter wildfire potential and fire season severity; the amount of change and the exact change at central locations worldwide have been obscure. They show that climate largely determines fire-prone regions. The climate characteristics of fire seasons were put into four classes, including Boreal, Temperate, Tropical, and Arid climate zones. Based on climate model projections, they determined how the fire-prone regions would change in extent and fire season length at the end of the twenty-first century. They concluded that due to global warming, the areas of the planet with frequent fire-prone conditions would increase by 29 percent, mainly in Boreal (+111 percent) and Temperate (+25 percent) zones, where there may also be a significant lengthening of the potential fire season. Their estimates of the expansion of fire-prone areas are uneven and more pronounced in the northern regions by a large percentage.

Stott, P. 2015. "Weather Risks in a Warming World." *Nature Climate Change* 5, no. 6 (April 27): 516–17. https://doi.org/10.1038/nclimate2640.

The paper by extreme weather and climate change expert Peter Stott discusses the impact of climate change on extreme weather events. It highlights that human activities currently drive 75 percent of daily heat extremes and 18 percent of heavy rain or snowfall events. The article emphasizes the need to limit global warming to mitigate these risks. Stott adds that further global warming will sharply increase the risks of extreme heat, rain, and snowfall. He clarifies that while climate change doesn't directly cause any single weather event, a warmer and moister atmosphere favors more frequent hot and wet extremes. Local variations in weather are already significant, and he states that if the world warms by 2 °C above preindustrial levels, human-caused climate change will drive 40 percent of rain and snow extremes and 96 percent of heat extremes. The probability of daily heat extremes would be twice as high in a 2 °C world compared to a 1.5 °C world.

Thompson, A. 2024. "July 2023 Is Hottest Month Ever Recorded on Earth." *Scientific American*, February 20, 2024. Accessed October 16, 2024. https://www.scientificamerican.com/article/july-2023-is-hottest-month-ever-recorded-on-earth.

Andrea Thompson, the associate editor covering the environment, energy, and Earth sciences for *Scientific American*, a senior writer covering climate science at Climate Central, and a reporter and editor at Live Science, where she primarily covered Earth science and the environment, published this article concerning the record-breaking temperatures in July of 2023. With a very warm start to July in 2023, the article reports the predictions by top climate agencies for that month to be the hottest month ever recorded on Earth—and likely the hottest in about 120,000 years. The article contains a graph published by NASA on July temperatures that would be useful to those interested in peak summer-time temperature trends. Because July is climatologically the hottest month of the year, July of 2023 was compared to paleoclimate evidence of past temperatures, including those found in ancient sediments and ice cores.

The article confirms the July 2023 record-breaking heat waves in the US Southwest, Mexico, China, and the Mediterranean. Thompson emphasizes the scientific link to rising global temperatures from the anthropogenic greenhouse effect. She also references the World Weather Attribution group, which found that the heat waves in North America and Europe were "virtually impossible" without climate change and that the heat waves in China were fifty times more likely to occur with current greenhouse gas concentrations. The article correctly highlights predictions by climate scientists like Karsten Haustein, a Leipzig University climate scientist who projected that 2023 would rank among the hottest on record and that El Niño combined with climate change boosts global temperatures. The World Meteorological Organization and C3S noted that July was about 1.5 °C (2.2 °F) above preindustrial levels. This was the limit the countries have agreed to strive for under the Paris Climate Accord. She stresses that although this does not mean the world has passed that point permanently, we must be concerned that the planet has warmed by about 1.2 °C (2.2 °F) since preindustrial times.

Wood, K., and O. Ashford. 2023. "The Ocean Can Play a Bigger Role in Fighting Climate Change Than Previously Thought." www.wri.org, September 2023. Accessed October 16, 2024. https://www.wri.org/insights/ocean-based-climate-change-solutions.

This article looks at ocean-based climate solutions that are currently ready and feasible. It gives information on the cost and amount of energy that could be generated by ocean-based renewable energy, reducing emissions from ocean-based transport, including commercial and industrial shipping, restoring blue-carbon ecosystems, expanding sustainable ocean food markets, increasing the absorption of greenhouse gases from the ocean, reducing offshore oil and gas, and overall investing in the oceans.

## Societal, Political, and Regional Impacts of Extreme Weather

### Books

Brewer, T. L. 2022. *Transforming U.S. Climate Change Policies*. SpringerBriefs in Energy, (Cham, Switzerland: Springer Nature). https://doi.org/10.1007/978-3-030-99716-8.

This work is a multidisciplinary analysis combining climate science, economics, technology, and political insights regarding US climate policy changes during 2021. It covers the climate crisis as of 2021, presenting key concepts, problems, and data in straightforward graphs and tables. It looks at the Biden administration Initiatives, which explored the wide-ranging decision-making during his first year in office, including state and local government policies. The book looks at the climate budget for fiscal year 2022, which revealed significant increases in funding for climate change programs and reductions in fossil fuel industry subsidies. It also examines critical industries such as energy, transportation, agriculture, and finance. The book also explores scenarios for addressing mitigation and adaptation challenges in the next decade, considering future climate changes. The book is a good reference for specific information on contemporary climate issues that will shape the future of policy in the United States.

Gardiner, S. M. 2011. *A Perfect Moral Storm: The Ethical Tragedy of Climate Change, Environmental Ethics and Science Policy Series.* Oxford Academic. Online edition, September 22, 2011. https://doi.org/10.1093/acprof:oso/9780195379440.001.0001.

This book takes on the ethical dilemma of climate justice. It argues that even though the public has known about climate change for decades, the current generation and those that came before us are still on a harmful path, including those not at fault for greenhouse emissions that cause global warming and destructive extreme weather events. The author's view is that this is an ethical problem in society. He challenges modern society, especially the wealthy, to see the harm they are inflicting with their choices on the global poor, future generations, and the natural world.

Incropera, F. P. 2016. *Climate Change: A Wicked Problem: Complexity and Uncertainty at the Intersection of Science, Economics, Politics, and Human Behavior*. Cambridge, UK: Cambridge University Press.

Frank P. Incropera's book delves into the multifaceted challenges posed by climate change. It explores the intricate connections between scientific understanding, economic considerations, political dynamics, and human behavior in addressing this urgent global issue. The book grapples with the complexities and uncertainties inherent in climate change, emphasizing the need for collaborative efforts across disciplines to find practical solutions. He discusses the disproportionate effects of climate-related events such as extreme weather, rising sea levels, and resource scarcity on marginalized populations. The book emphasizes the need for equitable solutions that consider social justice, resilience, and adaptation strategies to protect vulnerable communities from the adverse consequences of climate change.

Kalantzakos, S. 2017. *The EU, US and China Tackling Climate Change: Policies and Alliances for the Anthropocene*. Routledge Studies in Environmental Policy. Abingdon: Routledge.

Sophia Kalantzakos wrote this book shortly after the Paris Agreement came into effect. She looks at the possibility of a key relationship between the EU and China in terms of the goal to keep temperatures below 2 °C and other goals of the Paris Agreement. It examines why a close collaboration between the EU and China may result in the necessary push to solidify a vision and a roadmap for our common future in the Anthropocene. Kalantzakos introduces and looks at climate action leadership through an analysis of international relations. She argues that a close EU-China collaboration would be the best way to meet global climate change milestones. She proposes that since the EU and China do not compete for a global supremacy status, the collaboration between the two could be most efficient and rapid in the move toward a low-carbon future and dealing with the challenges of the Anthropocene. Kalantzakos demonstrates how Europe and China can exercise global leadership with climate change as a common interest and use their shared relations, development aid, and technology to formulate change. This book benefits students and scholars of environmental politics, international relations, climate change, and energy law and policy.

Mechler, R., L. M. Bouwer, T. Schinko, S. Surminski, and J. Linnerooth-Bayer. 2019. *Loss and Damage from Climate Change: Concepts, Methods and Policy Options*. Cham: Springer.

This book helps expose current limitations in loss and damage scholarly efforts and points to many needed advances in research where there are gaps and policy challenges ahead. It also provides a comprehensive and state-of-the-art overview of the field in its many offshoots driven by the cohort of researchers and practitioners from around the world, ranging from climatologists to lawyers, from anthropologists to macro economists.

## Articles and Reports

Associated Press. 2018. "Puerto Rico Issues New Data on Hurricane Maria Deaths." NBC News June 13, 2018. Accessed October 15, 2024. https://www.nbcnews.com/health/health-news/puerto-rico-issues-new-data-hurricane-maria-deaths-n882816.

This article looks at the death toll numbers from Hurricane Maria in 2017. The main argument in the article is that although extreme storm deaths are initially reported after the storm passes, the numbers do not often reflect the long-term effects of the storm that lead to additional deaths. Hundreds of people die due to post-storm conditions. Many lack water, electricity, or emergency care and fall to chronic diseases or infections. In particular, cases of sepsis, diabetes, and heart illnesses increased.

Aykut, S. C., and L. Maertens. 2021. “The Climatization of Global Politics: Introduction to the Special Issue.” *International Politics* 58: 501–18. http:doi.org/10.1057/s41311-021-00325-0.

This is an interesting look at climate change’s influence on geopolitics. The authors propose that climate change now constitutes a major defining goal or topic under many of the issues in world politics, so policy, economics, and social life are developing and changing under the scope of climate change. The concept of climatization refers to the frame of reference through which policy issues and forms of global activism are mediated and prioritized. This paper introduces the forces behind climatization and the “climate logic” that influence global politics.

Bailey, A., and L. Brush. 2022. “An Emerging Blueprint for Companies: Strategies to Advance Local Climate Resilience.” C2ES, September 2022. Accessed October 17, 2024. https://www.c2es.org/document/an-emerging-blueprint-for-companies-strategies-to-advance-local-climate-resilience.

Bailey and Brush dive into the effects of climate change on communities. They look into businesses and their climate resilience—the ability to anticipate, prepare for, and respond to hazardous events driven by anthropogenic climatic changes. They also discuss the need to supplement the 2021 Infrastructure Investment and Jobs Act and 2022 Inflation Reduction Act to meet the needs of communities that will still fall short even after funding from the two funds is enacted. Bailey and Brush discuss the need to solicit large companies and major stakeholders to build local climate resistance for their workers and anyone along their supply chain. They urge companies to invest in their communities.

Ballester, J., M. Quijal-Zamorano, R. Fernando Méndez Turrubiates, F. Pegenaute, F. R. Herrmann, J. M. Robine, X. Basagaña, C. Tonne, J. M. Antó, and H. Achebaket. 2023. “Heat-Related Mortality in Europe during the Summer of 2022.” *Nature News*, July 10, 2023. Accessed October 17, 2024. https://www.nature.com/articles/s41591-023-02419-z.

This highly referenced study published in *Nature* reports the findings on the 2022 summer heat waves in Europe that killed more than 61,000 people. Women were much more vulnerable to the heat than men, with 35,406 women perishing, compared to 21,667 men. The summer of 2022 was the continent's hottest on record. The authors analyzed total deaths as a result of heat using data from the Eurostat mortality database for thirty-five countries to estimate that 61,672 people died from heat-related illness between May 30 and September 4, with the highest deaths in Italy, Spain, and Germany. The article points out that Europe is still not as resilient to climate change as it needs to be despite the deadly 2003 heat wave that caused more than 70,000 excess deaths across the continent. The *Nature* study examined extreme heat's impact on specific countries and demographic groups. Italy had the most heat-attributable deaths throughout the summer, with 18,010, followed by Spain (11,324) and Germany (8,173).

While countries like France implemented heat prevention plans and other adaptation strategies to protect vulnerable populations, the study concluded that much of Europe is ill-prepared. Better efforts may have saved thousands in France as France had the highest temperature rise during the summer of 2022 at 4.4 °F, hotter than the 1991 to 2020 average. The authors warn that by 2030, the continent could see more than 68,000 heat-related deaths every summer. By 2050, each summer's deaths could top 120,000. The authors also emphasized the vulnerability of populations in Southern Europe and elderly women in general.

Cohen, S. 2022. "What Can We Do About Extreme Weather?" Columbia Climate School, Climate, Earth and Society, April 11, 2022. Accessed October 17, 2024. https://news.climate.columbia.edu/2022/04/11/what-can-we-do-about-extreme-weather.

This article looks at the public's concerns about extreme weather caused by climate change as the population who experience extreme weather events increases. While a large percent of the US population has experienced floods, fires, hurricanes, etc., due to an ever-increasing population and changes in land use, the frequency and magnitude of these events are helping the public be more curious about the science behind climate change and the connections to current weather events. This may help soften the partisan divides that were once knee-jerk reactions to climate science, climate policy, and increasing communal infrastructure. Many homes and industries are interconnected in their central water, sewage, communication, and electrical systems. In the aftermath of an extreme event, a collaborative and instantaneous response must be in place for the community to get back on its feet as quickly as possible. The article looks at the frequency of extreme weather events and the need to develop private and publicly subsidized insurance systems that pay the costs of reconstruction post-extreme events. The growing cost of insurance and taxes must be addressed, and it discusses the enactment of federally subsidized reconstruction insurance. The author contemplates the need for systems to reduce added homelessness, education disruption, and health care disruption.

Davies, A. R., V. Castán Broto, and S. Hügel. 2021. "Editorial: Is There a New Climate Politics?" *Politics and Governance* 9, no. 2 (July): 1–7. *Gale Academic OneFile.* https://doi.org/10.17645/pag.v9i2.4341.

This editorial looks at the emerging language and social nuances with the prioritization of climate change. It discusses the idea that climate politics to develop the Green new day is entering a new era. Highlighted in the editorial are the European Green Deal that hoped to create profoundly better climate policy by mainstreaming climate change into various sectors of policymaking and Biden's reintroduction of the United States into the Paris Agreement. And while China is using new industrial strategies with renewables, they point out loss opportunities for adaptation and mitigation in lower-income countries.

Dean, C. 2014. "Growing, and Growing Vulnerable: Barrier Islands Feeling the Effects of Climate Change." *New York Times*, September 29, 2014. Accessed October 17, 2024. http://www.nytimes.com/2014/09/30/science/barrier-islands-feeling-the-effects-of-climate-change-html.

Cornelia Dean looks at the imminent danger residents of barrier islands worldwide, particularly the Atlantic Ocean, face from rising sea levels. Dean dissects the report by the National Research Council to explain their findings. Among the report's key points is that rising sea levels are especially affecting barrier islands. At the same time, people are moving in masses to the coastal areas from Massachusetts to the Mexican border, where the strongest storms are hitting. The panel on the National Research Council calls for more action to manage coastline hazards. They advise limiting development on barrier islands to avoid billions in property damage as undeveloped lands recover more quickly from large storms.

Dunne, D. 2022. "Analysis: Africa's Unreported Extreme Weather in 2022 and Climate Change." Carbon Brief, October 26, 2022. Accessed October 17, 2024. https://www.carbonbrief.org/analysis-africas-unreported-extreme-weather-in-2022-and-climate-change.

The Carbon Brief website includes updated information on climate modeling, extreme weather, ice conditions, news from the IPCC, the state of the oceans, people, temperature, and global south climate database information. Carbon Brief is further described in Chapter 4 of this book. This particular article written by Daisy Dunne describes extreme events in Africa that scientists have attributed to global warming and climate change. Carbon Brief used disaster data, humanitarian group reports, and local interviews to track Africa's 2022 extreme weather events. The article points out that extreme weather events in Africa killed at least 4,000 people and affected 19 million since the start of 2022.

Falkenberg, M., A. Galeazzi, M. Torricelli, N. Di Marco, F. Larosa, M. Sas, A. Mekacher, W. Pearce, F. Zollo, W. Quattrociocchi, and A. Baronchelli. 2022. "Growing Polarization around Climate Change on Social Media." *Nature Climate Change* 12: 1114–21. https://doi.org/10.1038/s41558-022-01527-x.

This article looks at climate change and political polarization as two of the modern age's critical sociopolitical issues. The authors focus on the discussion around the United Nations Conference of the Parties on Climate Change (COP) using X data (then Twitter) from 2014 to 2021. They reveal a large increase in ideological polarization during COP26, following low polarization between COP20 and COP25. They find a 400 percent increase driven by growing right-wing activity since COP21 relative to pro-climate groups. They also identify a broad range of what they call "climate contrarian," which have become popular in Twitter climate discussions since 2019. These emphasized the theme of political hypocrisy views during COP26. The authors stress the need and importance of monitoring polarization and its impact on the public climate conversation.

Iberdrola. 2021. "¿Cuáles Son Los Países Más Amenazados y Vulnerables Por El Cambio Climático?" Iberdrola, April 22, 2021. Accessed October 17, 2024. https://www.iberdrola.com/sustainability/top-countries-most-affected-by-climate-change.

This article published by Iberdrola, a leading firm in clean energy production, looks at the top ten countries most vulnerable to climate change and the economic, social, and political reasons why. The article discusses the need for the amount world leaders to agree on regarding monetary funding and vulnerability scales for the countries listed.

Intergovernmental Panel on Climate Change (IPCC). 2022. "Poverty, Livelihoods and Sustainable Development." In *Climate Change 2022: Impacts, Adaptation, and Vulnerability. Contribution of Working Group II to the Sixth Assessment Report of the Intergovernmental Panel on Climate Change*, 1171–274. Cambridge, UK: Cambridge University Press. https://doi.org/10.1017/9781009325844.010.

Chapter eight of the IPCC's 6th assessment report in Working Group II's (WGII) *Climate Change 2022: Impacts, Adaptation, and Vulnerability* assesses the impacts of climate change on ecosystems, biodiversity, and human communities at regional and global levels. A special interest is given to vulnerabilities and the ability of humans to adapt to climate change. It examines societal inequalities that compound to make marginal communities the most vulnerable to rising temperatures, flood events, heat waves, and extreme storms. WGII stresses that vulnerable populations will be limited in their development, economic sustainability, and climate resilience in all emissions scenarios.

International Rescue Committee. "Early Needs Identification, Flood Disaster Sindh, Pakistan" September 6, 2022. Accessed October 17, 2024. https://reliefweb.int/report/pakistan/early-needs-identification-flood-disaster-sindh-pakistan.

This report published by OCHA, which focuses on identifying and conducting human relief efforts, looks at the vulnerability of communities in Sindh, Pakistan, in the wake of an extreme flooding event in 2022. They hope to identify early possible efforts for flood events. The ENI analysis identified the evolving basic needs of vulnerable populations dwelling in Sindh's flood-affected areas. Interviews of representatives of NGOs/INGOs/UN agencies, and government organizations were collected from the national and provincial disaster management authorities and other humanitarian agencies. This is a good reference for the considerations that must be taken in developing nations vulnerable to extreme weather events.

Ishfaq, H. M., and J. D. Ford. 2024. "Addressing the Climate Change Adaptation Gap: Key Themes and Future Directions." *Climate* 12, no. 2: 24. https://doi.org/10.3390/cli12020024.

This article is an extensive literature review and summary of the key issues facing the global finance community regarding aiding vulnerable nations and helping them adapt to climate change. It summarizes the challenges less financially stable nations face in adapting to the modern climate. It highlights the gaps in need versus the amount given and allocated. In particular, it looks at *The Adaptation*

*Gap Report 2023*, published by the United Nations Environment Program (UNEP), which examines the status of climate change adaptation efforts globally. The authors look at key themes in the report and incorporate an analysis of the wider literature and insights from COP28 to summarize key points and identify gaps where more work is needed to understand climate change adaptation. Their paper focuses on the underfinanced and underprepared state of global climate change adaptation efforts, the widening adaptation finance gap, slow progress in adaptation, gender equality and social inclusion issues, and challenges in addressing loss and damage. The authors also offer recommendations for future actions.

Ivers, L. 2022. "Social Dimensions of Climate Change." The World Bank, November 10, 2022. Accessed October 17, 2024. https://www.worldbank.org/en/topic/social-dimensions-of-climate-change.

This report by the World Bank looks at climate change and how it is related to global patterns of inequality. It emphasizes how the poorest and most vulnerable people suffer the brunt of climate change impacts yet contribute the least to the causes of global warming. The report looks at the millions of people facing unequal challenges in terms of extreme weather events, including female-headed households, the young and the old, persons with disabilities, Indigenous peoples and ethnic minorities, the homeless, social minorities, migrant workers, people in geographical locations where their financial, socioeconomic, cultural, and gender status is a threat and where their access to resources is impeded.

King, T., and J. Sterba. 2013. "C2ES Center for Climate and Energy Solutions." C2ES, July 23, 2013. Accessed October 17, 2024. https://www.c2Es.org/docment/u-s-companies-need-to-plan-for-extreme-weather-column.

Tom King, the US president of National Grid, an electrical and natural gas company serving seven million people in the US Northeast, and Jeff Sterba, the CEO of American Water, a public water utility company serving 14 million people in Canada, discuss their companies' efforts at climate resilience. After a record-warm year, they look at the rise of extreme weather events, from the flooding in Thailand in 2011 to Superstorm Sandy. Sterba reports that American Water is improving resilience to drought, flooding, and other extreme weather events by strengthening its infrastructure, establishing backup power supplies, promoting sustainable water use, and working with local and regional governments on disaster response. King offers a look into the efforts by National Grid. He lays out the efforts at ramping up climate resilience by assessing the company's vulnerability to extreme weather events, floodproofing critical equipment, having repair equipment and fuel at hand in case of a natural disaster, reducing the chance of downed power lines, and adding to the communication grid.

Mooney, A. 2023. "Developing Countries Need up to $387bn a Year to Adapt to Extreme Weather, Says UN." *Financial Times*, November 2, 2023. Accessed October 17, 2024. https://www.ft.com/content/d61425c5-0707-40bf-964b-4725a123d355.

This article in the *Financial Times* looks at the finance gap found by the United Nations Environmental Program when looking at how much money is needed by developing nations in harm's way of extreme weather. It also estimates how much money is available for funding. UNEP is reported to have declared a gap of $360 billion dollars in needed financial assistance for developing nations to build climate resilience in the face of extreme weather events.

Nutter, F. W. 1999. "Global Climate Change: Why U.S. Insurers Care." In *Weather and Climate Extremes*, edited by T. R. Karl, N. Nicholis, and A. Grahzi. Dordrecht, Springer. https://doi.org/10.1007/978-94-015-9265-9_6.

The link between natural weather events and the need for insurance coverage is undeniable. High-risk areas have expanded due to extreme weather, and the demand for insurance coverage can threaten insurance companies if the losses are too significant in a particular area. Those needing insurance often face high deductibles and high monthly costs, which generate a social injustice for those who cannot afford coverage. This book was one of the first in-depth looks at the economic losses and insured losses since the 1960s and its effects on those in extreme weather- high-risk areas. It discusses the US insurance industry's use of catastrophe models to integrate the knowledge of extreme events and actuarial science, which looks at risk probabilities. The author argued that insurance companies needed to realize the interdependence of insurance companies and climate and weather. It looks at the four times more significant increase in natural catastrophes from 1960 to 1999. Economic losses during that period were eight times greater than decades prior, and insured losses were fifteen times greater.

*Weather, Climate, and Society*. Journal. ISSN: 1948-8327; eISSN: 1948-8335.

*Weather, Climate, and Society* (WCAS) publishes research and reviews that address the many aspects discussed in this book, including economics, policy analysis, political science, history, communication, and institutional, social, health, and behavioral scholarship and research relating to weather and climate. The journal also houses articles about climate variability and longer-term climate change. Contributions include evidence-based analysis and substantive discussion of the interactions of weather and climate with society, taking an integrated approach and drawing on both the social and physical sciences.

Wilson, R. S., M. A. Méndez, E. Maibach, and G. Wong-Parodi. 2022. "Engaging Socially Vulnerable Communities and Communicating About Climate Change-Related Risks and Hazards." National Academy of Sciences. Accessed October 17, 2024. https://nap.nationalacademies.org/catalog/26734/engaging-socially-vulnerable-communities-and-communicating-about-climate-change-related-risks-and-hazards.

This report focuses on strategies for decision-makers to engage with socially vulnerable populations in the context of extreme weather events exacerbated by climate change. These vulnerable groups include pregnant individuals, children, low-income populations, Indigenous communities, immigrants, the elderly, people with disabilities, and those with preexisting medical conditions. By adopting effective communication and engagement approaches, decision-makers can empower these communities, address local concerns, and allocate necessary resources to adapt to climate-related hazards and risks. The report provides valuable insights into collaborative efforts among decision-makers, researchers, and communities.

## Websites and Interactive Tools

"Atlas of Mortality and Economic Losses from Weather, Climate and Water-related Hazards (1970–2021)." World Meteorological Organization, May 22, 2023. Accessed October 17, 2024. https://wmo.int/publication-series/atlas-of-mortality-and-economic-losses-from-weather-climate-and-water-related-hazards-1970-2021.

This exceptional website, created by the World Meteorological Organization (WMO), examines the global impacts of weather, climate, and water extremes between 1970 and 2021. The website includes a report and an interactive website featuring a story map where users can explore extreme weather's global and regional impacts by the number of deaths and economic loss. This study utilizes disaster data from the Emergency Events Database (EM-DAT). Disaster statistics are analyzed over fifty years and in decadal periods at national, regional, and global levels. A dedicated section highlights the disproportionate effects of tropical cyclones on disaster statistics, particularly in developing countries.

*Bulletin of the American Meteorological Society* (BAMS). Accessed October 17, 2024. https://www.ametsoc.org/index.cfm/ams/publications/bulletin-of-the-american-meteorological-society-bams.

The *Bulletin of the American Meteorological Society* is the flagship magazine of the American Meteorological Society. It publishes articles of interest and significance for the weather, water, and climate community. They publish news, editorials, and reviews. Those interested in climate change and extreme weather can find many of the latest articles on their site. Of specific interest is their special report—*The State of the Climate*, an annual peer-reviewed look at the world's climate health. Each summer in this BAMS Supplement, more than 400 scientists worldwide provide detailed updates on global climate indicators, newsworthy weather events, and other data collected on land, water, ice, and space. Their annual special report—*Explaining Extreme Events in the Face of Climate Change*, assesses the link between human-caused climate change and the strength and likelihood of extreme events for that year.

Cal-Adapt Climate Tools. Accessed October 17, 2024. https://cal-adapt.org/tools.

Cal-Adapt is an interactive website that features work sponsored by the California Energy Commission and the California Strategic Growth Council. It is a program developed under the state's climate change assessments by the Geospatial Innovation Facility at the University of California, Berkeley. The tool allows the public to explore often hard-to-read peer-reviewed research about climate change. It breaks down the data and gives critical information that summarizes the how state of California may be affected. The goal of this organization's research is to make climate data accessible. Eagle Rock Analytics, an environmental research and data analysis firm, leads the program. It receives funding from the California Energy Commission. Data contributors include UCLA, USGS, the California Department of Water Resources, Scripps Institution of Oceanography, and other environmental and educational organizations.

Center for Climate and Energy Solutions (C2ES). "Climate Basics: Extreme Weather Drought and Climate Change." Accessed October 17, 2024. https://www.c2es.org/content/drought-and-climate-change.

This website publishes updated maps of the current drought situation in the United States. The accompanying article discusses the latest trends in drought conditions within the United States. The site explains the scientific definition of drought and includes explanations of the effects of global warming on drought conditions. Warmer temperatures enhance evaporation, which reduces surface water and dries out soils and vegetation. Periods with low precipitation are drier than they would be in cooler conditions. It includes the understanding by climate scientists that climate change is altering the timing of water availability. The article explains the observation that warmer winter temperatures are causing less precipitation to fall as snow in the Northern Hemisphere, including in critical regions like the Sierra Nevada of California and the intermountain states that rely on snowpack for water management systems and for ecosystems that depend on the snowmelt. It mentions the positive feedback caused by the loss of precipitation in the form of snow that increases temperatures and further evaporation. An interesting finding is that some climate models show increased precipitation variability with projected times of extreme precipitation and drought. The decrease in consistent rainfall in the American Southwest is expected to continue. The website also estimates future seasonal and annual precipitation changes for locations that need further research. In general, scientists surmise that at the global scale, wet places will get wetter (those in latitudes where atmospheric cells generally form low-pressure systems), while relatively dry places (those in latitudes where atmospheric cells generally form high-pressure systems) will become drier.

Intergovernmental Panel on Climate Change (IPCC). 2021. "IPCC-WG1 Interactive Atlas." Accessed October 17, 2024. https://interactive-atlas.ipcc.ch.

The IPCC's Interactive Atlas is a state-of-the-art tool developed in the Sixth Assessment Report (AR6) by the Intergovernmental Panel on Climate Change (IPCC). Users of the Atlas can zoom in and out on world regions by periods for spatial and temporal analysis of documented and projected climate change information. Users can filter and select among different data sets, including temperature indices (mean temperature, maximum temperature, cooling and heating degree days, and more), precipitation, ocean sea surface temperatures, and other variables. The Atlas will then generate scenarios based on set emission targets. The visuals are dynamic and informative. The site includes a simple atlas for the public and an advanced atlas for researchers. Various data visualizations and summary statistics are available. The data displayed can be downloaded and shared.

Johannes, F., M. Ge, A. Pickens, and L. Vigna. 2023. "This Interactive Chart Shows Changes in the World's Top 10 Emitters." World Resources Institute, March 2, 2023. Accessed October 17, 2024. https://www.wri.org/insights/interactive-chart-shows-changes-worlds-top-10-emitters.

The World Resources Institute provides an interactive chart that explores changes in the world's top greenhouse gas emitters. The chart is updated every few years by the World Resources Institute based on the emissions of $CO_2$-equivalent (a carbon dioxide equivalent or $CO_2$ equivalent, abbreviated as $CO_2$-eq, is a metric measure used to compare the emissions from various greenhouse gases based on their global warming potential or GWP, by converting amounts of other gases to the equivalent amount of carbon dioxide with the same global warming potential). $CO_2$-eq, are commonly expressed as million metric tonnes of carbon dioxide equivalents. The chart also lists the breakdown of emissions by sector for the top ten emitters. Along with the chart is an analysis of the current top ten emitters.

National Aeronautics and Space Administration (NASA). "Fire Information for Management Systems." Accessed October 17, 2024. https://firms.modaps.eosdis.nasa.gov/map/#d:24hrs;@5.4,0.0,2.6z.

NASA manages an interactive map that visualizes fire occurrence data obtained from their satellites (Landsat, MODIS, and VIRS) over twenty-four hours, present day, seven days, and historical (selection of date required). Additional layers can be added to the map, such as protected land areas, volcanic activity, road infrastructure, and human settlements. This website offers a fantastic resource for those interested in fire science, fire event trends, and management of fire sectors.

National Geographic Extreme Weather Challenge Game. Accessed October 17, 2024. https://education.nationalgeographic.org/resource/extreme-weather.

National Geographic developed an interactive game for grades four to ten called Challenge: Extreme Weather! The game shows students exciting careers in science and engineering while displaying essential concepts in engineering design. Those interacting with the challenge also learn about the layers of the atmosphere and the different types of instruments used to measure weather data. Students can take on the role of an engineer or a meteorologist at Weather COMM, where they can simulate the launching of weather balloons to forecast a thunderstorm, use programming to save researchers in a winter storm, test a weather-resistant structure against hurricane forces, and design an aircraft to fight a forest fire and weather obstacles. While the connection between climate change and extreme weather and climate change is not the focus, students will learn the tools and methods used by many meteorologists who collect data to assess the link between current conditions and past norms.

National Oceanic and Atmospheric Administration. "Science and Information for a Climate Smart Nation." Accessed October 17, 2024. https://www.climate.gov.

The United States' National Oceanic and Atmospheric Administration climate.gov website offers national and regional information on land and ocean information in the form of reports, news, blogs, maps, interactive graphs, and data. The site also offers teaching resources.

## Documentaries and Media

Cooper, D., and Z. Canepari, dir. 2019. *Fire in Paradise*. Netflix. Accessed October 17, 2024. https://www.netflix.com/title/81050375.

This forty-minute documentary film premiered at the Telluride Film Festival and won an Audience Award at the Hamptons International Film Festival. It documents the devastating wildfire that ravaged Paradise, California, in 2018. The film details the experiences of residents who survived the Camp Fire in Butte County, California, which killed eighty-five people and caused $16.5 billion in damage. The event is recreated through interviews with the courageous Cal Fire and volunteer firefighters, 911 fire dispatch operators, and residents—all trapped in a wall of flames. With the addition of news coverage and live phone footage from panicked callers to rescue operations, the film brings to reality the devastation of a wildfire consuming a community. The film drives the message of the increasingly frequent and intensified wildfires California is projected to experience with continued global warming.

Elizondo, G. 2017. "Hurricane Maria Turns Dominica into 'Giant Debris Field.'" Aljazeera, September 25, 2017. Accessed October 17, 2024. https://www.aljazeera.com/videos/2017/9/25/hurricane-maria-turns-dominica-into-giant-debris-field.

This video report by Gabriel Elizondo, an Aljazeera reporter, recorded the devastation caused by flood damage in Dominica when Hurricane Maria hit the island. It is an ethnographic capture of the feelings of those struggling after the destruction. One can see the damage to infrastructure, powerlines, supermarkets, and hospitals. It looks at the fact that the small, poorer islands of the Caribbean are at the forefront of extreme weather threats.

Hamilton, D., dir. 2018. *Decoding the Weather Machine.* NOVA.https://utah.pbslearningmedia.org/collection/nova-decoding-the-weather-machine/.

This two-hour documentary examines extreme weather events from disastrous hurricanes, extensive droughts and wildfires, heat waves, and floods. It looks at the connection between a changing climate and the response that short-term weather events express to the increased heat in the atmosphere. NOVA uses scientific testimony from climate experts to cut through the confusion around climate change. It explains why scientists overwhelmingly agree that our climate is changing and human activity is causing it. It also discusses whether the trajectory of increased extreme weather events in our future can be curtailed. Scientists from around the world offer an optimistic view of mitigation strategies that humans can incorporate to change the trajectory and thrive in the future.

PBS. 2012. *Earth: The Operator's Manual.* Accessed October 17, 2024. https://www.pbs.org/newshour/show/earth-the-operator-s-manual-chronicles-conservation.

This three-hour documentary series by the Public Broadcast Systems was produced in 2012. Dr. Richard Alley, a leading authority on climate and geology, narrates the series. The documentary discusses topics still of interest to those hoping to understand the most pressing components of the climate system and extreme weather events. Specifically, the documentary showcases topics such as the human need for energy and the use of fossil fuels, how $CO_2$ affects the climate, how scientists use ice cores to study past climate shifts, the US government's view on climate change (2012), efforts by the US military and other nations to reduce the use of fossil fuels, and innovative ways to provide society with clean energy.

The documentary was accompanied by a book of the same title, providing evidence of human-caused global warming. It also offered various mitigation and adaptation strategies. The book discusses the rise of human-generated use of fossil fuels to power everyday activities and the consequences that modern-day conveniences have brought upon the Earth's systems. Using one engaging story after another, coupled with accessible scientific facts and world authority on climate change, Dr. Richard Alley explores the fascinating history of energy use by humans over the centuries, giving proof that already-high levels of $CO_2$ were already causing damaging global warming.

## Podcasts

Conn, A. 2019. "Not Cool Ep 10: Stephanie Herring on Extreme Weather Events and Climate Change Attribution." Not Cool. Podcast audio, October 1, 2019. Accessed October 17, 2024. https://futureoflife.org/podcast/not-cool-ep-10-stephanie-herring-on-extreme-weather-events-and-climate-change-attribution.

In the tenth episode of "Not Cool," the host delves into the connection between climate change and extreme weather events. Stephanie Herring, a National Oceanic and Atmospheric Administration climate scientist, joins her. Stephanie's work on extreme event attribution has earned her a spot on *Foreign Policy* magazine's list of Top 100 Global Thinkers. During their discussion, they explore shifts in attribution research, worrisome trends, the significance of data in decision-making, and more. Herring discusses that climate change, driven by human-induced greenhouse gas emissions, intensifies extreme weather events, making them more destructive and frequent. Topics she covers are wildfires in the western United States, extreme rainfall in China, and droughts in South Africa, which have all been impacted by climate change.

Lawler, J. 2021. "Climate Now Introduction." Climate Now. Podcast audio, March 31, 2021. Accessed October 17, 2024. https://climatenow.com/podcast/climate-now-introduction.

Founded by James Lawler, Climate Now hosts experts on climate change and energy systems. Funded by the Lawrence Livermore Foundation, Climate Now communicates the science of the changing climate, the technologies that could help us avoid a climate crisis, and the economic and policy pathways to achieve net zero emissions globally. The website features a podcast with over 150 interviews on topics from the state of Atlantic hurricanes to harvesting energy from algae. The website also offers videos and regional workshops on climate change and energy solutions.

PBS. 2022. "The Science of Extreme Weather." WHYY PBS NPR. Podcast audio, November 11, 2022. Accessed October 17, 2024. https://whyy.org/episodes/the-science-of-extreme-weather.

This podcast offers a look at the way that scientists are actively researching not only how to predict extreme weather more accurately but also how to effectively communicate their warnings to the public. The episode also looks at the aftermath of storms, exploring how communities recover and prepare for future weather events. Some intriguing stories include a storm-chasers mission to understand tornadoes, the resolution of a meteorological debate during the 1993 blizzard, and the impact of severe weather on patient's access to life-saving treatments.

7

# Chronology

The following is a chronological list of some of the defining moments and major events impacting the development of global warming and climate change science, and the link of global warming to extreme weather.

**1750–1870** First Industrial Revolution begins as coal, petroleum, and other fossil fuels are used for machinery and fuel. Land clearing also speeds up greenhouse gas emission. Population growth is seen from improvements in agriculture and medicine.

**1824** Joseph Fourier, French mathematician and scientist, calculates that the Earth would be far colder if it lacked an atmosphere to insulate the Earth. This is the first proposal of what becomes known as the greenhouse effect.

**1856** Eunice Foote, American scientist, becomes the first to conclude that certain gases warm when exposed to sunlight, and that rising carbon dioxide ($CO_2$) levels could change atmospheric temperature and could affect climate.

**1859** John Tyndall, Irish scientist, proves the connection of atmospheric $CO_2$ and the greenhouse effect. His study of infrared radiation and the physical properties of air demonstrates that some gases block infrared radiation, and notes that changes in the concentration of the gases could bring climate change.

---

Highway 50 in South Lake Tahoe, California, during a blizzard in March 2024. An atmospheric river dumped an extreme amount of snowfall in the Sierra Nevada mountains. Alpine meadows recorded 80 inches of snow, while South Lake Tahoe received 51 inches. (Claire White/Dreamstime.com)

**1879** Global weather data is compiled and standardized by the International Meteorological Organization. This includes global temperatures, wind, storms, etc.

**1896** Svante Arrhenius, Swedish Nobel Laureate chemist, publishes first calculation of global warming from human emissions of $CO_2$.

**1897** T. C. Chamberlin, American glaciologist, studies the ice ages and produces a model for global carbon exchange including feedbacks.

**1920–5** Texas and Persian Gulf oil fields open and begin the time of cheap energy.

**1930s** Milutin Milankovitch, Serbian scientist, proposes the relationship and position of the Earth as the cause of the ice ages every 100,000 years.

**1938** Guy Steward Callendar, English steam engineer and inventor, reads "The Artificial Production of Carbon Dioxide and Its Influence on Temperature" before the Royal Society. He explains that: (1) humans increased the amount of $CO_2$ in the atmosphere by burning fossil fuels; (2) increased $CO_2$ could cause an increase in global temperatures; and (3) global temperatures were already rising. He develops Arrhenius's theory and shows that the land temperature of Earth rose over the previous fifty years before 1938.

**1945** US Office of Naval Research begins funding scientific studies including those of the physical properties of the ocean, the atmosphere, and the biosphere. Many of these studies contribute to the understanding of the interconnectedness of Earth's systems that apply to climate change.

**1955** Norman Phillips, American Meteorologist, generates the first general circulation model or GCM of monthly and seasonal tropospheric patterns. The models become the bedrock of weather and climate studies.

Improvements in infrared instrumentation allow for very precise measurements of atmospheric $CO_2$.

**1956** Wallace Broecker, Maurice Ewing, and Bruce Heezen gather evidence from deep-sea and lake deposits, that a global climate shift of as much as 5–10 °C can take place in less than 1,000 years if feedback loops.

Maurice Ewing and William Donn build an elaborate model for the coming and going of ice ages based on the idea that a retreat of snow and ice would bring more warming by sunlight. Their new idea was that feedback involving ocean currents triggered by the melting of ice sheets contributes to sea level rise. The warm water then spills into the Arctic Ocean and melts the ice cover and speeds up the warming. The study also concludes that once the Arctic Ocean is free of ice so much moisture evaporates producing heavy snowfall and reversing the warming trend back to cooling. An ice-free world could trigger the next ice age.

The National Center for Atmospheric Research (NCAR), initially known as the National Institute for Atmospheric Research (NIAR), is a leading research facility established in Boulder, Colorado, under the management of the University Corporation for Atmospheric Research (UCAR). NCAR is heavily involved in climate change research through the development of advanced climate models, such as the Community Earth System Model (CESM), which aid in predicting climate conditions. The center's work encompasses studies on climate extremes, geoengineering, sea level rise, and heat waves. Additionally, NCAR's projects receive grants to enhance climate projections, thereby providing crucial information for policymakers to address and mitigate the impacts of climate change.

**1958** Roger Revelle establishes the first Committee on Climate Change and the Ocean (CCCO) under the Scientific Committee on Ocean Research (SCOR) and the International Oceanic Commission (IOC). He directs the main center for the Atmospheric Carbon Dioxide Program at University of California, San Diego.

Charles Keeling, American scientist, accurately measures $CO_2$ in the Earth's atmosphere and detects an annual rise. The $CO_2$ level is 315 ppm, and the mean global temperature (five-year average) is 13.9 °C. He is able to set up $CO_2$ monitoring stations at Mauna Loa, Hawaii, and elsewhere to create a set of global baseline measurements of carbon dioxide. The "Keeling Curve," the upward-sloping, saw-toothed graph of atmospheric $CO_2$, begins to show yearly increases from anthropogenic sources.

**1963** First meeting of experts concerned with global warming warns that a rise in sea level is likely, with "immense flooding" of shorelines.

**1965** Edward Norton Lorenz, American mathematician and scientist, joins others at a meeting in Boulder, Colorado on causes of climate change and points out the chaotic nature of the climate system and the possibility of sudden shifts.

**1967** The International Global Atmospheric Research Program is established and begins to gather data for better short-range weather prediction, and longer-term climate trends.

Syukuro Manabe, a Japanese computer scientist, manages a team to build a mathematical simulation of the planet's atmosphere. Manabe's simulation quantifies, for the first time, the relationship between $CO_2$ and the temperature of Earth's atmosphere. Manabe's simulation provides evidence that: (1) doubling the atmospheric concentration of $CO_2$ could cause Earth's average surface temperature to rise by more than 4 °F and this can raise ocean levels by a hundred feet; (2) his simulation predicts that $CO_2$ traps heat energy in the lower atmosphere. The Earth's surface and its oceans therefore get hotter, while the upper atmosphere would cool. This combination—cooler above, hotter below—is regarded by climatologists as the smoking gun of human-caused climate change because other explanations such as the sun growing brighter, would uniformly heat the atmosphere at every altitude.

**1968** Extended droughts, failed monsoons, and untimely rains upset global markets and led to increases in domestic food prices.

**1969** M. Budyko and W. Sellers independently introduce low-order climate models that continue to play an important role in the mathematical modeling of climate. The models investigate the role ice-albedo feedback plays in influencing surface temperature.

Nimbus III satellite begins to provide comprehensive global atmospheric temperature measurements.

**1970** Creation of US National Oceanic and Atmospheric Administration (NOAA), the world's leading funder of weather, ocean, and climate research.

**1971** MIT sponsors the study of man's influence on physical systems. The Study of Man's Impact on Climate (SMIC) in Stockholm, Sweden, brought together some of the world's top atmospheric and environmental scientists to discuss the many ways in which human activities, including the addition of $CO_2$ from the burning of fossil fuels, might affect global atmospheric processes.

**1972** Ice core studies show climate variability in the past.

Droughts in Africa, Ukraine, India cause world food crisis. Fears spread about climate change.

**1974** Cooling from human pollution by aerosols suspected to be as likely as warming. This is the beginning of the public's confusion about the mechanisms of climate change. Some researchers predict a cooling of the planet due to a projected increase of aerosols that cool the planet. J. Murray Mitchell concludes "Of the two forms of pollution, it appears that the carbon dioxide increase is more influential in raising planetary temperatures than the anthropogenic particle increase is in lowering planetary temperatures."

**1976** Deep-sea core sediment studies show a dominating influence from 100,000-year Milankovitch orbital changes, supporting the role of feedbacks.

Deforestation and other ecosystem changes contributing to the addition of carbon are recognized as major factors in the future of the climate.

**1977** Scientific consensus on global warming, not cooling, begins as the chief climate risk in the next century.

**1978** The National Climate Program Act of 1978 was passed by US Congress over concerns about the environmental impacts of $CO_2$-induced climatic change, weather modification, international droughts and food security as well as the availability of reliable forecasting information to meet domestic agricultural needs. Many climate scientists welcomed the act insofar as it promised to streamline climate information.

The American Association for the Advancement of Science (AAAS), included climate change as a potential subject of focus.

**1979** World Climate Research Program launched to coordinate international research on climate and weather.

**1980** US President Ronald Reagan's appointees sought to dismantle environmental programs, downplaying environmental risks across the board. The new administration cut social science and environmental science research, including an 87 percent cut to the budget of the Department of Energy's new Solar Energy Research Institute in Boulder, Colorado.

**1981** Al Gore, then representative from Tennessee, confronts Reagan's Department of Energy appointee, N. Douglas Pewitt, on the administration's efforts to slash the federal budget for $CO_2$ research which had been approved under the National Climate Policy Act.

James Hansen, American scientist, and others, show that sulfate aerosols can significantly cool the climate, raising confidence in models that incorporate aerosols and show future greenhouse warming.

1981 becomes the warmest year on record since 1850.

**1982** Greenland ice cores reveal drastic temperature shifts possible within 100 years as evident in the distant past.

**1983** Greenhouse warming becomes part of political conflict in the US. Reports from US National Academy of Sciences and Environmental Protection Agency sparked conflict.

**1985** V. Ramanathan and other authors propose that global warming may come twice as fast as expected considering the contributions of methane and other trace greenhouse gases.

The United Nations sponsors the Villach Conference where 100 scientists meet and begin talks on climate change much like the COP meetings of today. Consensus is reached among experts that some global warming seems inevitable. The scientists in attendance call on governments to consider international agreements to restrict greenhouse gas emissions.

**1987** Heat waves and droughts in the eastern United States, India, and Russia rekindled many of the climate-related food security concerns voiced in the mid-1970s. Global warming began to emerge more frequently and more sensationally in the national media.

**1988** The International Toronto Conference on the Changing Atmosphere is held with 300 scientists attending. Those in attendance agree that the world was going into an era of unprecedented climate change. Toronto conference calls for strict, specific limits on greenhouse gas emissions.

UK Prime Minister Margaret Thatcher is the first major leader to call for action on greenhouse gas emissions.

The United Nations and the World Meteorological Organization establish the Intergovernmental Panel on Climate Change (IPCC).

**1989** The Global Climate Coalition is established by fossil-fuel and other US industries to persuade politicians and the public that climate science is too uncertain to justify action.

**1990** The first IPCC report (first assessment report or FAR) says the climate has been warming and future warming seems likely. The report highlighted the importance of climate change as a global challenge with global consequences and one that would require global collaboration to address and mitigate.

The United Nations Framework Convention on Climate Change (UNFCCC, or "U-N-F-Triple-C"), the key international treaty to reduce global warming and cope with the consequences of climate change, was established.

**1991** Global warming skeptics proposed that the temperature changes observed in the twentieth century were primarily due to solar influences such as sunspots and solar radiation, and were responsible for the warming trend. However, by the first decade of the twenty-first century this theory was largely discredited as further research showed that the increase in global temperatures could not be explained by solar activity alone.

Geologists discover massive extinction event fifty-five million years ago, caused by an increase in greenhouse gases that lead to global warming.

**1992** The United Nations Conference on Environment and Development (UNCED) in Rio de Janeiro, Brazil, also known as Earth Summit, Rio, introduces a variety of treaties and proposals to promote an inclusive, internationalist vision of sustainable development, including the United Nations Framework Convention on Climate Change.

The UNFCCC becomes the first international legal treaty written specifically to deal with the problem of climate change.

**1995** The second IPCC assessment report (SAR) detects "fingerprint" of human-caused greenhouse effect warming and predicts serious warming in the coming century.

The coverage of the breakup of the Antarctic ice shelves and other signs of actual current warming in polar regions begin showing evidence to the public that global warming is causing effects in polar regions.

**1997** The International Panel on Climate Change convenes in Kyoto, Japan. The meeting results in the Kyoto Protocol, setting targets for industrialized nations to reduce greenhouse gas emissions. The US does not sign any treaties to reduce greenhouse gas emissions.

**1998** A "Super El Niño" contributes to an exceptionally warm year.

Global Circulation Models reproduce current climate.

The Pew Center on Global Climate Change and its Business Environmental Leadership Council is established to provide a new voice for effective, pragmatic climate solutions.

**1999** Michael E. Mann, climatologist and geophysicist, produces the "hockey stick" temperature data indicating current warming is historically unprecedented.

**2000** Oil lobbyists convince the US administration to ignore the warning from scientists about greenhouse gases and global warming.

**2001** The third IPCC report (TAR) concludes that unprecedented global warming is "very likely," with highly damaging future impacts. It also focused attention on the impacts of climate change and the need for adaptation.

**2002** Studies find surprisingly strong "global dimming," due to pollution, has retarded arrival of greenhouse warming, but dimming is now decreasing.

**2003** Scientists publish several observations raising concern over the collapse of the West Antarctica and Greenland ice sheet possibly raising sea level faster than previously projected.

Europe suffers a deadly summer heat wave.

Senators John McCain and Joseph Lieberman produce the first bipartisan bill introduced in Congress for a comprehensive market-based climate solution.

**2005** The Kyoto treaty goes into effect. Major industrial nations except the US begin work to reduce emissions including Japan, Western Europe, and some US regional governments and corporations.

The European Union initiates "cap-and-trade" market to reduce emissions.

Hurricane Katrina and other major tropical storms cause the beginning of the discussion on storms intensifying due to global warming.

C2ES—Climate Dialogue at Pocantico, occurs in which governments, businesses, and civil societies from around the world convene to produce key concepts reflected in United Nation agreements.

**2006** Al Gore's "An Inconvenient Truth" documentary is viewed by many highlighting the Keeling Curve and the influence of fossil fuels on global warming.

California passes a pioneering cap-and-trade program.

China overtakes the United States as the world's largest $CO_2$ emitter.

**2007** The fourth IPCC report (AR4) warns that serious effects of warming are evident and that the cost to ecosystems and society would be far greater than the cost of reducing emissions. It also laid the groundwork for a post-Kyoto agreement, focusing on limiting warming to 2 °C.

The United States' Supreme Court decides in favor of the State of Massachusetts, alongside eleven other states, several US cities, and a number of private organizations in their bid to petition the US Environmental Protection Agency to regulate $CO_2$ and other greenhouse gases as pollutants.

Thirty major companies and nonprofits in the US Climate Action Partnership help advance bipartisan action in Congress.

**2008** Climate scientists warn that even with halting greenhouse gas emissions immediately, global temperature will remain elevated for millennia.

**2009** Climate scientists report that damage from climate change and extreme weather events is arriving at a faster pace than was anticipated just a few years earlier.

Copenhagen IPCC COP meeting fails to negotiate binding agreements for greenhouse gas emissions. Scientists are concerned about limited hope for avoiding future severe weather events and climate disasters.

**2010** C2ES publishes "Adapting to Climate Change," the first in a series of initiatives to help communities and companies strengthen their climate resilience.

**2011** The Pew Center officially becomes the Center for Climate and Energy Solutions, pursuing the same mission with a new name and new supporters.

The Climate Registry, and Environmental Protection Agency organize the first annual Climate Leadership Conference and Awards.

Carbon Capture Coalition is established to successfully advocate for federal support for carbon capture.

**2012** The Doha Amendment to the Kyoto Protocol was extended for a second commitment period, starting in 2013 and lasting until 2020. The amendment included a revised list of greenhouse gases to be reported on by parties in the second commitment period; and amendments to several articles of the Kyoto Protocol.

Attribution studies find some recent disastrous heat waves, droughts, extremes of precipitation, and floods were made worse by global warming.

**2014** C2ES launches the "Toward 2015" iniative. Dialogue among lead negotiators from key countries opens, ultimately producing an essential blueprint for the historic Paris Agreement.

**2015** The COP 21 in Paris results in the Paris Agreement, a pledge by nearly all nations to set trajectories for greenhouse gas cuts and to report their progress.

Researchers report that the collapse of the West Antarctic ice sheet may be irreversible, bringing meters of sea level rise and increased flooding over future centuries.

C2ES launches the Solutions Forum to bring together businesses, cities, and states around clean energy, climate resilience, and greenhouse gas reduction.

**2016** The US Conference of Mayors is established in the Alliance for a Sustainable Future, helping cities and businesses collaborate on climate action.

**2017** Twenty-five major companies publish full page ads in *The New York Times*, *Wall Street Journal* and *New York Post* with signatures from companies including Adobe, Apple, Facebook, Google, Hewlett Packard Enterprise, Intel, Microsoft, Morgan Stanley, Salesforce, Tiffany & Co., and Unilever urging President Trump to stay in the Paris Agreement.

**2018** The fifth IPCC report on 1.5 °C warming says that to avoid dangerous climate change, the world's greenhouse gas emissions must be in sharp decline by 2030.

**2019** Increasing extreme weather disasters including tropical cyclones, wildfires, and drought promote demonstrations and civil disobedience such as strikes.

Climate Innovation 2050 releases Pathways to 2050. The plan includes scenarios for decarbonizing the US economy, working with companies and a team of outside experts to visualize and test different pathways for reducing US emissions by 80 percent by midcentury.

Getting to Zero is released, a US Climate Agenda, a comprehensive new agenda outlining the policies needed over the coming decade to put the United States on the path to carbon neutrality by 2050.

**2021** The sixth IPCC report warns that catastrophic outcomes including a continued increase of extreme weather events cannot be ruled out.

**2022** Democrats in US Congress pass legislation providing subsidies to reduce the US's greenhouse gas emissions.

Mean global temperature is 14.8 °C, the warmest in tens of thousands of years. The level of $CO_2$ in the atmosphere is 421 ppm, the highest in millions of years.

President Biden passes the Inflation Reduction Act of 2022, the single-largest federal investment in climate and clean energy passed to date.

# Glossary

**Adaptation:** In human systems, the process of adjustment to actual or expected climate and its effects, in order to moderate harm or exploit beneficial opportunities. In natural systems, the process of adjustment to actual climate and its effects; human intervention may facilitate adjustment to expected climate and its effects.

**Albedo:** The proportion of sunlight (solar radiation) reflected by a surface or object, often expressed as a percentage.

**Anthropogenic:** Referring to environmental change caused or influenced by people, either directly or indirectly.

**Atmosphere:** The gaseous envelope surrounding the Earth, divided into five layers: the troposphere, which contains half of the Earth's atmosphere, the stratosphere, the mesosphere, the thermosphere and the exosphere, which is the outer limit of the atmosphere.

**Attribution Science:** The science that uses the process of evaluating the relative contributions of multiple causal factors to a change or event with an assessment of confidence.

**Biomass:** Organic material excluding the material that is fossilized or embedded in geological formations.

**Bipartisan:** Involving the agreement or cooperation of two political parties that usually oppose each other's policies.

**Blue Energy:** Covers all technologies to harvest the renewable energy of our seas and oceans other than offshore wind.

**Cap-and-Trade:** A system for controlling carbon emissions and other forms of atmospheric pollution by which an upper limit is set on the amount a given business or other organization may produce but which allows further capacity to be bought from other organizations that have not used their full allowance.

**Carbon Capture:** Capturing carbon dioxide ($CO_2$) at emission sources, transporting and then storing or burying it in a suitable deep, underground location.

**Carbon Cycle:** The process that moves carbon between plants, animals, and microbes; minerals in the earth; and the atmosphere.

**Carbon Dioxide:** A colorless, odorless gas produced by burning carbon and organic compounds and by respiration. It is naturally present in air (about 0.03 percent) and is absorbed by plants in photosynthesis.

**Carbon Negative:** Emit less than zero carbon dioxide and carbon dioxide equivalent ($CO_2$-eq) greenhouse gases.

**Carbon Neutral:** Making no net release of carbon dioxide to the atmosphere, especially through offsetting emissions by planting trees.

**Carbon Offsets:** An action intended to compensate for the emission of carbon dioxide into the atmosphere as a result of industrial or other human activity, especially when quantified and traded as part of a commercial program.

**Carbon Positive:** That activity goes beyond achieving net-zero carbon emissions to create an environmental benefit by removing additional carbon dioxide from the atmosphere.

**Carbon Tax:** The government sets a price that emitters must pay for each ton of greenhouse gas emissions they emit.

**Clausius–Clapeyron Relationship:** In chemical thermodynamics specifies the temperature dependence of pressure, most importantly vapor pressure, at a discontinuous phase transition between two phases of matter of a single constituent. Its relevance to meteorology and climatology is the increase of the water-holding capacity of the atmosphere by about 7 percent for every 1 °C (1.8 °F) rise in temperature.

**Clean Energy:** Energy that comes from renewable, zero emission sources that do not pollute the atmosphere when generated and used, as well as energy saved by energy efficiency measures.

**Climate:** Consistent averages of weather conditions over a long period of time.

**Climate Change:** A systematic change of average climate conditions, sustained over several decades or longer.

**Climate Ethics:** Concerns what is morally right or wrong (fair or unfair, just or unjust) when it comes to making decisions about climate change.

**Climate Models:** Use of mathematical equations to characterize how energy and matter interact in different parts of the ocean, atmosphere, land.

**Climate Resilience:** The ability to anticipate, prepare for, and respond to hazardous events driven by anthropogenic climate changes.

**Climatization:** Points to a powerful yet uneven social process in which climate change is increasingly becoming the frame of reference for the mediation and hierarchization of other global issues.

**Confirmation Bias:** The tendency to interpret new evidence as confirmation of one's existing beliefs or theories.

**Constituents:** A member of a body of voters in a specified area who elect a representative to a legislative body.

**Cyclone:** A system of winds rotating inward to an area of low atmospheric pressure, with a counterclockwise (Northern Hemisphere) or clockwise (Southern Hemisphere) circulation; a depression. Also called a hurricane in the Atlantic, or a typhoon in the Pacific.

**Developing Nation:** A sovereign state with a less developed industrial base and a lower Human Development Index (HDI) relative to other countries.

**Downpour:** A heavy rainfall.

**Drawdown:** The future point in time when levels of greenhouse gas concentrations in the atmosphere stop climbing and start to steadily decline.

**Drought:** A deficiency of precipitation over an extended period of time (usually a season or more), resulting in a water shortage. Indicators of drought include precipitation, temperature, streamflow, ground and reservoir water levels, soil moisture, and snowpack.

**Earth's Energy Budget:** Describes the balance between the radiant energy that reaches Earth from the sun and the energy that flows from Earth back out to space.

**Egalitarian:** Relating to or believing in the principle that all people are equal and deserve equal rights and opportunities.

**Emissions:** Substances released into the air and are measured by their concentrations, or parts per million, in the atmosphere.

**ENSO (El Niño Southern Oscillation):** A warming of the ocean surface, or above-average sea surface temperatures (SST), in the central and eastern tropical Pacific Ocean.

**Equilibrium Climate Sensitivity:** The long-term temperature rise (equilibrium global mean near-surface air temperature) that is expected to result from a doubling of the atmospheric $CO_2$ concentration ($\Delta T2\times$).

**The G7:** The Group of Seven (G7) is an intergovernmental political & economic forum consisting of Canada, France, Germany, Italy, Japan, the United Kingdom, and the United States; additionally, the European Union (EU) is a "non-enumerated member." It is organized around shared values of pluralism, liberal democracy, and representative government.

**Geo-engineering:** Describes a range of ways to intervene on a large scale in the Earth's natural systems—the oceans, soils and atmosphere—to directly combat climate change.

**Global Change Research Act of 1990:** A comprehensive and integrated United States research program that will assist the nation and the world to understand, assess, predict, and respond to human-induced and natural processes of global change.

**Green Energy:** Energy that can be produced using a method, and from a source, that causes no harm to the natural environment.

**Green Hydrogen:** Green hydrogen is essentially hydrogen produced by splitting water into hydrogen and oxygen using renewable electricity.

**Greenhouse Gases:** A gas that contributes to the greenhouse effect by absorbing infrared radiation, e.g., carbon dioxide and chlorofluorocarbons.

**Heatwave:** A period of abnormally hot weather, often defined with reference to a relative temperature threshold, lasting from two days to months.

**Hurricane:** A tropical storm with winds that have reached a constant speed of seventy-four miles per hour or more.

**Industrial Revolution:** The process of change from an agrarian and handicraft economy to one dominated by industry and machine manufacturing.

**Industrialization:** The development of industries in a country or region on a wide scale.

**Intergovernmental Panel on Climate Change (IPCC):** The United Nations body for assessing the science related to climate change.

**Inundation:** A flood, or the fact of being flooded with water.

**Keeling Curve:** A graph that shows the ongoing change in the concentration of carbon dioxide in Earth's atmosphere.

**Marginalized Groups:** Peoples or populations are groups and communities that experience discrimination and exclusion (social, political, and economic) because of unequal power relationships across economic, political, social, and cultural dimensions.

**Methane:** A colorless, odorless flammable gas which is the main constituent of natural gas. It is the simplest member of the alkane series of hydrocarbons. It is also a very powerful greenhouse gas.

**Mitigation:** Involves reducing the flow of heat-trapping greenhouse gases into the atmosphere, either by reducing sources of these gases (for example, the burning of fossil fuels for electricity, heat, or transport) or enhancing the "sinks" that accumulate and store these gases.

**Nationally Determined Contributions:** A climate action plan to cut emissions and adapt to climate impacts.

**The ND-GAIN Index:** Ranks 181 countries using a score which calculates a country's vulnerability to climate change and other global challenges as well as their readiness to improve resilience. The more vulnerable a country is the lower their score, while the more ready a country is to improve its resilience the higher it will be.

**Net Zero:** Refers to the balance between the amount of greenhouse gas (GHG) that's produced and the amount that's removed from the atmosphere.

**New Blue Economy:** A sustainable and equitable ocean and coastal economy that optimizes advances in science and technology to create value-added, data-driven economic opportunities and solutions to pressing societal needs.

**Nitrous Oxide:** A colorless gas with a sweetish odor, prepared by heating ammonium nitrate.

**Nomadic/Semi-nomadic:** A people living usually in portable or temporary dwellings and practicing seasonal migration but having a base camp at which some crops are cultivated.

**Pacific Decadal Oscillation (PDO):** A long-lived El Niño-like pattern of Pacific climate variability.

**Pandemic:** A worldwide outbreak of a disease in humans in numbers clearly in excess of normal.

**Pastoralist:** People who practice pastoralism as a livelihood system. Pastoralism is the extensive livestock production system that involves the tracking and use of grazing and water across a given landscape (normally a "rangeland"). Normally practiced in dryland areas, mobility is key to this system.

**Photosynthesis:** The process by which green plants and some other organisms use sunlight to synthesize foods from carbon dioxide and water.

**Polar Vortex:** A very large, long-lived, rotating low-pressure system located near the north or south pole, especially in winter.

**Radiation:** The emission of energy as electromagnetic waves or as moving subatomic particles, especially high-energy particles which cause ionization.

**Representative Concentration Pathway (RCP):** a greenhouse gas concentration (not emissions) trajectory adopted by the IPCC.

**Saffir Simpson Scale:** A 1 to 5 rating based only on a hurricane's maximum sustained wind speed.

**Shared Socioeconomic Pathway (SSP):** Provide narratives describing alternative socioeconomic developments.

**Solar Radiation:** General term for the electromagnetic radiation emitted by the sun.

**Supply Chains:** The sequence of processes involved in the production and distribution of a commodity.

**Sustainability:** Involves ensuring the persistence of natural and human systems, implying the continuous functioning of ecosystems, the conservation of high biodiversity, the recycling of natural resources and, in the human sector, successful application of justice and equity.

**Typhoon:** A tropical storm in the region of the Indian or western Pacific oceans.

**Vulnerable Populations to Climate Change:** Low-income households, communities of color, those experiencing homelessness, and immigrant populations. Many of these groups live in urban areas that are prone to extreme heat (urban heat islands). Some may also live in housing that does not have adequate insulation or cooling.

# Index

**Bold** page numbers indicate tables, *italic* numbers indicate figures.

# About the Author

**Mariangelica (Maria) Groves** holds a B.S. in Earth Science from The Pennsylvania State University (1997) with a focus on Geographic Information Systems (GIS), environmental management, meteorology, and climate change. Her senior thesis was mentored by Nobel Prize winner for his work with the intergovernmental panel on climate change, Dr. Richard Alley. Under his supervision, she worked on ice cores from the South Cascade Glaciers. In 2009 she earned a Master's degree in instructional design and technology. An emphasis was placed on using learning theories and applying them to create online, modular instruction. Her doctoral degree in Geography from the University of Utah focused on assessing climate change during the late Holocene (past 3,000 years) caused by natural as well as anthropogenic disturbances using fossilized pollen, charcoal, and the geochemistry from sediment cores.

She has taught at Utah Valley University, the University of Utah and is currently at Weber State University in Ogden, Utah. Her teaching load includes courses in climate change, meteorology, arid lands, national parks, and field courses in the southeastern desert of Utah. Her research interests include climate change on the Colorado Plateau, fire ecology, and regenerative farming practices in arid lands. She lives in Sandy, Utah with her husband and two daughters and two very fuzzy dogs who love to accompany her on hikes in the mountains. When not working you can find her on the ski slopes or traveling.